T0146224

# The Natural Philosophy
of Margaret Cavendish

Here on this Figure Cast a Glance,
But so as if it were by Chance,
Your eyes not fixt, they must not stay,
Since this like Shadowes to the Day
It only represent's; for Still,
Her Beuty's found beyond the Skill
Of the best Paynter, to Imbrace,
Those louely Lines within her face,
View her Soul's Picture, Iudgment, witt,
Then read those Lines which Shee hath writt,
By Phancy's Pencill drawne alone
Which Peece but Shee, Can iustly owne.

# THE NATURAL PHILOSOPHY
# OF MARGARET CAVENDISH

*Reason and Fancy during the Scientific Revolution*

LISA T. SARASOHN

The Johns Hopkins University Press
*Baltimore*

The Johns Hopkins University Press
2715 North Charles Street
Baltimore, Maryland 21218-4363
www.press.jhu.edu

Library of Congress Cataloging-in-Publication Data

Sarasohn, Lisa T., 1950–
The natural philosophy of Margaret Cavendish : reason and fancy
during the scientific revolution / Lisa T. Sarasohn.
p. cm.—(The Johns Hopkins University studies in historical
and political science)
Includes bibliographical references and index.
ISBN-13: 978-0-8018-9443-5 (hardcover : alk. paper)
ISBN-10: 0-8018-9443-3 (hardcover : alk. paper)
1. Feminism and science—England—History—17th century.
2. Science—Philosophy—History—17th century.  3. Newcastle,
Margaret Cavendish, Duchess of, 1624?–1674—Knowledge—
Physics.  I. Title.
Q130.S266 2010
509.42'09032—dc22          2009026921

A catalog record for this book is available from the British Library.

*Frontispiece.* From Margaret Cavendish, *Grounds of Natural Philosophy*,
second edition, London: A. Maxwell, 1668. Drawing by artist Abraham
Jansz Van Dieppenbeck and engraving by Pierre Louis van Schupp.
Courtesy The Newberry Library.

*Special discounts are available for bulk purchases of this book. For more
information, please contact Special Sales at 410-516-6936 or
specialsales@press.jhu.edu.*

The Johns Hopkins University Press uses environmentally friendly
book materials, including recycled text paper that is composed of at
least 30 percent post-consumer waste, whenever possible. All of our
book papers are acid-free, and our jackets and covers are printed
on paper with recycled content.

# Contents

# Acknowledgments

This book began as a footnote to my study of Pierre Gassendi. Margaret Cavendish was almost unknown in 1984 when I published "A Science Turned Upside Down: Feminism and the Natural Philosophy of Margaret Cavendish, Duchess of Newcastle," *Huntington Library Quarterly* 47 (1984): 289-307. In the decades since, both Cavendish studies and gender history have grown in importance and forced a reexamination of seventeenth-century intellectual life. I wish to thank those who have contributed to my understanding of Cavendish's natural philosophy, by both critiquing my work directly and inspiring me to ask questions that did not occur to me when I began my work on this extraordinary woman.

In particular, I benefited from the comments and suggestions of Mary Terrall, Mary Baine Campbell, Paula Findlen, James Fitzmaurice, Brandie Siegfried, Sara Mendelson, Stephen Clucas, Hilda Smith, William Newman, Bernard Lightman, and Lawrence Principe. The advice of Margaret J. Osler has been, as always, invaluable. Steven Shapin has helped shape my thinking about the early modern scientific community. The work of Sarah Hutton and Michael Hunter has provided much needed context for the history of the Royal Society and the role of women in natural philosophy in the seventeenth century. Katharine Park's work on wonders and popular culture helped broaden my understanding of Cavendish. Although she probably doesn't remember, Carolyn Merchant's early response to a paper I delivered on Cavendish also set me thinking in different directions. All errors in this book are, of course, my own.

My colleagues in the Oregon State History Department works-in-progress group have patiently listened to much of this book, in various incarnations, and it is richer for their comments. I would like to thank them all: Paul Farber, Mary Jo Nye, Robert Nye, Jeff Sklansky, Mina Carson, Ben Mutschler, William Husband, Marisa Chappell, Kara Ritzhaber, Maureen Healy, Jon Katz, Nicole von Germeten, and Paul Kopperman. In addition, I profited from the advice of Gary Ferngren and Anita Guerrini. I was very fortunate that during my time at Oregon

State a doctoral program in the History of Science was established through an endowment by Thomas Hart Horning and Mary Jones Horning. This program brings eminent scholars to OSU and creates an invigorating climate for the development of ideas about the history and culture of science.

I also want to thank my colleagues in the International Margaret Cavendish Society. Every two years, scholars from many disciplines and many places come together to share their insights. It is a society remarkable for its cohesiveness and civility. The many different approaches to Cavendish's work have enriched my own. In addition to Brandie Siegfried, James Fitzmaurice, and Sara Mendelson, the current, former and future presidents of the society, I would like to thank Gweno Williams, Emma Rees, and Erna Kelly.

Support for my project was provided by a Mellon Fellowship in the Humanities, a William Andrews Clark Short-Term Fellowship, a Folger Library Short-term Fellowship, an Oregon State College of Liberal Arts Research Fellowship, an OSU Library Travel grant, and a fellowship at the Center for the Humanities at OSU. Librarians at Chatsworth House, U.K., the British Library, the Bodleian Library, the Folger Library, and the Clark Library have been invariably helpful. At OSU, Elissa Curcio has helped me greatly in all matters Cavendish.

Parts of this study have appeared in different guises in "Margaret Cavendish and Patronage," *Endeavour* 23 (1999): 130–32; "*Leviathan* and the Lady: Cavendish's Critique of Hobbes in the *Philosophical Letters*," in *Authorial Conquests: Essays on Genre in the Writings of Margaret Cavendish*, edited by Line Cottegnies and Nancy Weitz (Madison, NJ: Fairleigh Dickinson Press, 2003; London: Associated Universities Presses, 2003), 40–58; and "Métaphysique et Mab: Le Premier Atomisme de Margaret Cavendish (1623–1673)," in *Gassendi et la modernité*, edited by Sylvie Taussig (Turnhout, Belgium: Brepols, 2008).

I have presented papers on Cavendish at the History of Science Society, the Renaissance Society of America, and at the UCLA Working Historians Group. I am particularly grateful to the unknown questioner who asked me at the History of Science meeting in Cambridge in 2003 who the audience was for the joke Cavendish played on the Royal Society. Much of my work since then has been based on that question.

The editors at the Johns Hopkins University Press have been exacting and encouraging. I wish to thank Robert J. Brugger and Josh Tong for helping me to clarify my thinking and keeping me on task. I apologize for all uses of the passive voice in this acknowledgment and the book. My copy editor, Grace Carino, has been more than gracious about uncertain spelling, both Cavendish's and mine.

Last of all, I wish to thank my students and my family. Both have put up with endless monologues about a seventeenth-century woman in whom they had but a passing interest. However, all were more than pleased to hear that the book was finally going to see the light of day. My students in my honors Western Civilization classes and my Women's History class were excited to find out what a professor does when she does her own work. It was a reminder that the scholarly world professors inhabit is cradled in a university where learning is a universal goal.

Margaret Cavendish said that she was her "lords scholar" and her book was her "childe." I have also been blessed with a husband who is unfailingly supportive, as well as being a terrific writer and editor. This book would not exist without him. And, happily, the book is not my only child. I dedicate this book to my sons, Alex and Peter.

# The Natural Philosophy
# of Margaret Cavendish

# Gender, Nature, and
# Natural Philosophy

The emblematic device of the Royal Society, the mark of its belief in experimentation as a way to understand nature, was the air-pump, used by Robert Boyle
and Robert Hooke to measure the pressure of air and the viability of living creatures in a vacuum. But the air-pump generated a result the experimenters did not
anticipate, at least for one imaginative commentator. "Then came the Lice-men,
and endeavoured to measure all things to a hairs breadth, and weigh them to an
Atome," wrote Margaret Cavendish in her satiric account of the new science, *The
Description of a New World, called The Blazing World.* "But," she continued,
"their weights would seldom agree, especially in the weighing of Air, which they
found a task impossible to be done; at which the Empress began to be displeased,
and told them there was neither Truth nor Justice in their Profession; and so dissolved their society."[1]

Other beastly scientists inhabit the fantastic world created by Cavendish in her
1666 satire. "We take more delight in Artificial Delusions, then [*sic*] in natural
truths," declare the bear-men experimental philosophers. If their telescopes and
microscopes actually found the truth, these monstrous scientists argue, "we should
want the aim and pleasure of our endeavours in confuting and contradicting each
other; neither would one man be thought wiser then [*sic*] another, but all would
either be alike knowing and wise, or all would be fools."[2] Ape-men "chymists"
foolishly waste their time trying to find the philosopher's stone. The Empress of
the Blazing World, voice and avatar of Cavendish herself, dissolves all their societies, hoping to avoid the factions that might bring "an utter ruine upon a State
or Government."[3]

In *Blazing World*, published six years after the foundation of the Royal Society,
Cavendish used parody to figuratively demolish an institution she viewed as dangerous, useless, and deluded in thinking that its experimental program could ri-

val and confine the works of nature. Historically, the Royal Society is widely seen as a part of the resolution of the upheavals in religion, science, and politics that shook Europe and England in the seventeenth century.[4] Cavendish saw it differently.

In many different genres—essays, treatises, poetry, romance, orations, plays, and treatises—Cavendish brought intellect and awareness to analyzing the implications of the new science for nature and women. Historians lament the absence of female voices in the past, but sometimes a single voice reveals common experiences and concerns. Commenting on the changes she observed during a tumultuous personal and public life, Cavendish criticized mechanical and experimental philosophy, Aristotelianism, Epicureanism, and alchemy. As the first woman to publish her own natural philosophy, Cavendish was completely singular—and strange—in her time and place, but her singularity reveals the revolutionary potential of many of the ideas and practices she questioned. Margaret Cavendish had a scientific revolution, but it was different from the one that has made it into the history books.

Cavendish adapted the ontology of the mechanistic philosophers—matter in motion—but reconfigured it to suit her own view of nature. Her natural philosophy, a form of vitalistic materialism that posited a universe composed of three kinds of matter—rational, sensitive, and inanimate matter—is no stranger to modern definitions of science than are the philosophies of many of the thinkers Cavendish critiqued in her works, including Hobbes, Descartes, Henry More, and Joan Baptista Van Helmont. Her work is dialogic and polemical. It is also anachronistic, at least in terms of form. Boyle and other members of the experimental community rejected romance, poetry, and speculative fiction as legitimate means for expressing philosophic ideas about nature. Cavendish embraced all genres as a vehicle for her ideas and had no compunction about injecting philosophic discussion into the middle of comedies of manners or utopian fiction. It may be that she used different rhetorical tactics to insinuate her ideas into a public forum increasingly closed to women in the late seventeenth century, but she may also reflect the discursive practices of earlier natural philosophers, like the Italian naturalists and alchemists who presented their ideas in many different forms. Among other seventeenth-century English thinkers, Francis Bacon had used a scientific fantasy, *The New Atlantis*, to express his plan for scientific organization, and Thomas Hobbes had employed a kind of natural fantasy, the state of nature, as the grounds for constructing his political philosophy.

Cavendish learned the basics of the new philosophies from her husband, William Cavendish (1593–1676), the Earl (1628), Marquis (1643), and Duke of Newcastle (1665; referred to here as Newcastle), and her brother-in-law, Sir Charles Cavendish (1595–1654), who exposed her to the diverse currents of early modern intellectual thought. The earl and Margaret Cavendish met while she was a lady-in-waiting to Henrietta Maria at the English court in exile in Paris, where they married in 1645.[5] Active patrons and participants in science in the first half of the seventeenth century, the earl and his brother encouraged Cavendish's interest in all things philosophic.

Newcastle was a patron of Thomas Hobbes, also at the court of Henrietta Maria, and Newcastle and Sir Charles entertained Hobbes, René Descartes, and Pierre Gassendi. Both brothers were part of a European-wide epistolary network whereby new ideas were presented and debated. They discussed the new philosophies with Cavendish, who decided that the radical questioning of earlier ideas and epistemologies entitled her to contribute to scientific discourse and discovery. Skepticism, in particular, about the ability of anyone to find demonstrative truth prompted her scientific speculations. "Who knows?" she asks repeatedly in her works, and her answers reflected a new way for women to participate in articulating scientific ideas. Cavendish's natural philosophy represents a gendered reading of the new science in which all of nature is invested with life, knowledge, and feeling and each kind of material being—human, animal, vegetable, and mineral—has its own worth.

It is not that there was something particularly female or womanly about Cavendish's natural philosophy, as if gender means some ontologically universal state of being. "Gender" is a malleable category, possessing no objective meaning outside the society in which it is constructed.[6] It consists of the characteristics culturally associated with male and female—or even the definition of female and male—at any given time. When Cavendish wrote a gendered natural philosophy, she reflected the traits associated with women and men in the seventeenth century, both positive and negative. Some men and women viewed women as nurturing, imaginative, docile, chaste; others characterized them as proud, stubborn, talkative, and lascivious.[7] Most men and women viewed patriarchy as the natural state of the human family. In Cavendish's universe, there is usually an absent male figure, who has handed authority over to a female. Thus, in Blazing World, the Emperor gives the Empress total power after marrying her; in another tale published in 1654, "The She-Anchoret," the eponymous heroine's father dies after providing her with education and financial independence. In Cavendish's natural philosophy, God gives the ordering of the world to a female Nature and is rarely seen again. In Cavendish's own life, her widowed mother guided and controlled the family estates for her children, just as Elizabeth I had governed the realm she inherited from her father. In all these instances of female power, however, authority originated with the male. As singular and original as Cavendish was, and although the transferred power once received becomes almost absolute, she could not imagine an original source of female power.

Similarly, Cavendish in her early works envisioned her reason and fancy as receptive to the imprints of "knowledge and understanding, wit and the purity of language" conveyed by her husband. "I am my Lords Scholar," she writes, borrowing the material of her fancies from her husband before reconstructing them with her own imagination.[8] Thus, Cavendish's singularity developed within the parameters of the most traditional of social institutions: marriage. Newcastle championed his wife, even when she challenged the boundaries of social and intellectual propriety.

Newcastle had been a spectacularly unsuccessful general during the English Civil War but was more successful as a patron of the arts and sciences.[9] He was himself a writer of plays, poetry, and treatises on horsemanship and swordsmanship, which brought him more renown than his efforts as a military man.[10] He and his brother, Sir Charles Cavendish, had pursued natural philosophy from at least 1630.[11] While in Paris in 1647, Hobbes, Descartes, and the atomist Pierre Gassendi met in a kind of intellectual salon at Newcastle's residence in Paris. Cavendish was sometimes present at these meetings, but she disclaimed any direct interaction with the philosophers, who spoke in French and Latin, while she claimed to know only English.[12] The Cavendish brothers became the "patrons" of her intellectual activities. In defense of Cavendish's early works, Newcastle wrote in 1655, "But here's the crime, a Lady writes them, and to intrench so much upon the male prerogative, is not to be forgiven."[13]

Many recent feminist scholars ask, implicitly or explicitly, "Did women have a scientific revolution?"[14] Standing in the center of this question—and its answer—is Margaret Cavendish, the only woman to publish her works on natural philosophy in the seventeenth century. According to the traditional account of the scientific revolution, science was both a cause of and a solution to the crisis of authority in philosophy, part of the uncertainty and anxiety of the early seventeenth century under the dual assaults of new religions and new ideas.[15] In the late seventeenth century, order was reintroduced when investigators of nature no longer sought to understand the why of things but merely how they work. According to this view, the scientific revolution began with Copernicus and culminated in the genius of Newton, the transcendent seer and scientist who divided the 1600s, allowing its more enlightened decades to become part of the long eighteenth century.

Today, historians attach considerable reservations to this triumphalist account. Students of the scientific revolution, in particular, criticize this narrative's omissions: the varied sources of modern science, from alchemy to magic; the differing contexts for the development of ideas; the contributions of craftsmen; the preoccupation with physics and astronomy at the expense of chemistry, medicine, and natural philosophy.[16]

And, in an omission more noticeable now than in the past, women.

Scholars have sought to repair this gap, both by emphasizing the previously ignored role of women in the scientific revolution and by analyzing the effect of gender on their thought. Recent explorations of the role of English women in science cite the many women participating in knowledge production in the seventeenth century. It was not inevitable that women would be excluded from the emerging scientific community. In the first half of the century, women were part of epistolary networks of scholars, participated in salons, and were sometimes members of guilds. The household was not separated from the laboratory, and the practice of chemistry and medicine was local and included female practitioners. Cavendish's first forays into scientific writing in the 1650s may have reflected this informal and developing scientific community, to which her husband and brother-in-law belonged. In this context, their encouragement of her writing becomes more understandable.

Some scholars believe that there was enough flux in the intellectual community to allow the participation of women until the 1650s, or indeed the 1730s. Much of this argument rests on different locations: England or the Continent, cities or the countryside. In other words, the historical role of women—and Cavendish's role in particular—has to be contextualized in terms of precise circumstances. The strikingly active and public role played by reformist and religious women during the civil war and Commonwealth periods made such participation more problematic after the Restoration of Charles II in 1660. The demarcation of the household, the marketplace, and the state became increasingly distinct after the 1660s—although the "domestication" of women was not complete until the nineteenth century.

Cavendish ran into trouble in the 1660s when she tried to join the public discourse of the scientific community. Her exclusion from the formal institutions of science, particularly the Royal Society, shows how the scientific revolution (if we allow that term to persist) drove women from direct involvement in scientific activity. As a more distinct community of investigators of nature slowly emerged, association with women may have been seen as diminishing claims to authority and status in science, as it did in politics. In European history, the more law was defined and institutionalized, the more women found their rights confined. In the enforcement of authority, formerly undefined behavior became suspect or unacceptable. A similar development characterizes the scientific revolution.

After her return to England in 1660, the same year as the foundation of the Royal Society, Cavendish's agenda became much more polemical. She defined her natural philosophy in contrast to the views of the mechanical philosophers and alchemists and especially the new experimental philosophy. Interestingly, the duke, who was somewhat involuntarily retired from public life, never joined the Royal Society—although several of his clients and his cousin William Cavendish, the Earl of Devonshire, did.[17] One of those clients, Walter Charleton, arranged for the duchess to visit the Royal Society in 1667, an event much discussed at the time and in modern literature about the early days of the society.[18]

The Royal Society tried to adhere to Francis Bacon's charge never to claim more than could be known through systematic investigation of matters of fact. The idea of coherence in science may have been the impetus Cavendish needed to systematize her own thought, which she did in several treatises published during the 1660s. Apologizing for her early works, Cavendish wrote at the beginning of her *Philosophical Letters* (1664), "I cannot say, I divulged my opinions as soon as I had conceiv'd them, but yet I divulged them too soon to have them artificial and methodical."[19] As the possibilities of entering any intellectual community lessened, Cavendish became increasingly vehement in defining and explaining her scientific ideas, often to the point of endless repetition. Against a wall of dismissal and derision, she argued that "Worthy Authors, were they my censurers, would not deny me the liberty they take themselves." But while her male contemporaries used method to restrain their claims to knowledge, Cavendish used her gender to claim unlimited freedom to write what she wished: "And if I should express more Vanity then Wit, more Ignorance then Knowledg, more Folly then Discretion, it being according to the Nature of our Sex."[20]

Cavendish used her gender as both explanation and apologia for her writings. In a dedication in *Observations*, she wrote, "But as for Learning, that I am not versed in it, no body, I hope, will blame me for it, since it is sufficiently known, that our Sex is not bred up to it, as being not suffer'd to be instructed in Schools and Universities." But, somewhat disingenuously given her previous claims to instruction from her husband and others, she concludes, "I will not say, but many of our Sex may have as much wit, and be as capable of Learning as well as Men, but since they want Instructions, it is not possible they should attain to it." Ultimately, however, female ignorance may be a good thing, "for Learning is Artificial, but Wit is natural."[21]

Sometimes female insight could trump male intellect. In the preface to the 1663 edition of *Philosophical and Physical Opinions*, Cavendish denigrated the rhetoric of the educated natural philosophers, and by implication, she emphasized the insight of her own scientific ideas: "their Artificial Arguments being as Clouds which Obscure the Natural Light of Information or Observation, for there is as much Difference between Logistical Arguments, and Natural Observations, as between Light and Darkness, and the best Natural Philosophers are those, that have the Clearest Natural Observation, and the Least Artificial Learning."[22]

In Cavendish's lexicon, natural is better than artificial, and consequently an unlettered woman is the best kind of natural philosopher. But Cavendish believed herself exceptional in overcoming intellectual exclusion. She was anxious when she saw the effects of contemporary social conventions on women themselves. In one of her earliest works, she urged professors and students to value her work and to "receive it without a scorn, for the good encouragement of our sex, lest in time we should grow as irrational as idiots, by the dejectednesse of our spirits, through the carelesse neglects, and the despisements of the masculine sex to the effeminate, thinking it impossible we should have either learning or understanding, wit or judgement, as if we had not rational souls as well as men, and we out of a custom of dejectednesse think so too, which makes us quit all industry towards profitable knowledge being imployed onely in looe, and pettie imployments."[23]

Because they are denied education and not allowed to develop their higher faculties, Cavendish argued, women become bestial and irrational: "We are become like worms that onely live in the dull earth of ignorance." Here is a monstrous change indeed; categories are blurred, and unnatural hybrids are created. "What ever did we do," asks Cavendish, "but like Apes, by Imitation?" The oppression of men "hath so dejected our spirits, as we are become so stupid, that Beasts are but a Degree below us, and Men use us but a Degree above Beasts."[24]

The result of the dehumanization of women, according to Cavendish, is that most women have indeed become stupid, caring for nothing but frippery and fancy: "Neither doth our Sex delight or understand philosophy, for as for Natural Philosophy, they study no more of Nature's works than their own Faces . . . and for a Moral Philosophy they think that too tedious to learn, and too rigid to practice."[25] Sounding more like a Puritan preacher than a social critic, Cavendish indicts most women for embracing the role society has defined for them. Thus, her singularity is not only a tool in establishing her position as a natural philosopher but also a statement of moral rectitude. Cavendish recognized that gender characteristics were the result of cultural practices and beliefs, but she nevertheless often condemned women for adhering to such norms.

Cavendish was in a particularly good place to recognize the constraints that structured female potentialities. She was the pampered youngest daughter of a gentry family, the Lucases, who were led by an indomitable widowed mother, Elizabeth Leighton Lucas. Cavendish claimed that she had little formal education. In her autobiography, she wrote that while she and her sisters were tutored in "singing, dancing, playing on Musick, reading, writing, working, and the like, yet we were not kept strictly thereto, they were rather for formalitie than benefit." Her mother "cared not so much for our dancing and fidling, singing and prating severall languages; as that we should be bred virtuously."[26]

Cavendish's mother thought that virtue was more important than accomplishments or education. In her earliest works, Cavendish adhered to the viewpoint, derived from Aristotle and commonly held by the educated classes, that women's brains, composed of moist and cold matter, were naturally inferior to men's brains.[27] Nevertheless, even while accepting this physiological inferiority, Cavendish argued that some women could, with proper cultivation, become more intelligent than some men: "And though it seem to be natural, that generally all Women are weaker than Men, both in Body and Understanding . . . yet some are far wiser than some Men; like Earth; for some Ground, though it be Barren by Nature, yet, being well mucked and well manured, may bear plentifull Crops, and sprout forth divers sort of Flowers. . . . So Women by education may come to be far more knowing and learned than some Rustick and Rude-bred men."[28] Hence, through education earthly women rise above earthly peasants, who many elite thinkers believed did not possess the capacity for reason and in fact were bestial. The cultivation of women, here figured like the cultivation of the earth with which women were associated, allows them to transcend boundaries and even, Cavendish continues, rule.[29]

Moreover, in some instances not even nurture is necessary, for Nature has provided the elevation of the female gender. Cavendish herself demonstrates the potential superiority of (a) woman. Although Nature has not made women as strong or rational as men, she has made women from her "finer parts . . . which seems as if Nature had made Women as purer white Manchet, for her own Table, and Palat, where Men are like coarse household Bread which the servants feed on."[30] In this rather cannibalistic analysis of the hierarchy of gender, women are the white bread of Lady Nature, while men are the brown bread fit only for Nature's servants.

From a material point of view, which increasingly will become Cavendish's outlook, the mix of matter and form could produce a woman as wise as some men, or, in a radical inversion of traditional hierarchy, a woman could possess qualities that would even elevate her above coarse and common males. Such women would be particularly adept at imaginative and emotional "sweet conceits . . . which makes them nearest to resemble Angells, which are the perfectest of all her [Nature's] work." Paradoxically, the one female superiority Cavendish could claim because of her sex, the fine white bread of fancy, would stick in the craws of male natural philosophers. She was cognizant of the lack of palatability of her imaginative productions, but she used education and nature as the warrant for her speculations.

In her 1662 play "Youths Glory, and Deaths Banquet," Cavendish created a learned avatar of herself, the Lady Sanspareille, who lectures on natural philosophy but at first is considered monstrous by philosophers. "O times! O manners," one proclaims, while another stutters, "Beauty and favour and tender years, a female which nature hath denied hair on her Chin, so smooth her brow, as not to admit one Philosophycall wrinckle, and she to teach, a Monster tis in Nature; since Nature hath denyed that sex the fortitude of brain." But after hearing her, the philosophers proclaim, "We will now all send for Barbers, and in our great Philosophies despair, shave of our reverend beards, as excrements . . . and stuff boyes footballs with them."[31]

The Lady has unmanned the philosophers, modern and ancient; old Aristotle himself would "wish he had never been the master of all Schooles, now to be taught, and by a girle." But the affront to men is not completely unexpected: Nature, like Lady Sanspareille, is a woman; one of the philosophers concedes: "Nature, those dost us wrong, and art too prodigall to the effeminate Sex; but I forgive thee, for thou art a she, dame Nature thou art."[32]

## KNOWING NATURE

Cavendish's concept of nature, more than any other aspect of her natural philosophy, separated her from the rest of the scientific community in the late seventeenth century—even from those who included some principle of vitality in their concepts of matter. Cavendish reformulated the definition of nature in her works, mingling anthropomorphized metaphors of nature as a woman with a depiction of matter as internally self-moving and self-conscious.

Traditionally in premodern Europe, nature was personified as a female. But the seventeenth century radically reexamined the nature of nature. More than a century ago, the sociologist Max Weber argued that nature was "disenchanted" in the sixteenth and seventeenth centuries. In the early modern period, all spiritual and animistic forces—what the historian of science Lorraine Daston has recently characterized as "preternatural"—were drained from the natural world. In premodernity, Daston notes, "the preternatural embraced strange weather, figured stones, petrifying springs, the occult virtues of minerals, and the myriad other deviations from the ordinary course of nature. Marvels were not so much violations of as much as exceptions to the natural order."[33]

Describing the prevailing attitude, Robert Boyle could write, "Men so often have recourse to nature, and think she must extraordinarily interpose to bring such things about." But it is "God's agency," he argues, that brings such "anomalies" to pass. God, not needing an intervening nature, does all himself.[34] Instead of a natural world infused with soul, in the late seventeenth century nature was increasingly perceived as a Cartesian machine operating according to natural laws instituted by God. The mechanical philosophy embraced by some members of the Royal Society seemed to render matter itself inert and lifeless. Wrote Boyle, "Matter alone, unless it is moved, is altogether inactive," and when natural philosophers "tell us of such indeterminate agents as the soul of the world, the universal spirit, the plastic power, and the like . . . they tell us nothing, that will satisfy the curiosity of an inquisitive person."[35] Boyle felt that experimenters revealed the glory of God when they investigated his handiwork and denied nature an active role in the ordering of the universe, either as a force operating on matter or as an internalized principle of being.

But Cavendish thought that nature was full of activity, both as a personified female semidivine regent, like the Empress of the Blazing World, and as the entirety of vitalized matter constituting the universe and its parts. In her view, neither aspect of matter was accessible to the experimental philosophers, whose experiments attempted both too much and too little. On the one hand, they thought that they could penetrate nature and reveal her secrets; on the other, they thought that the only approach to nature was through experiment, not reason. To Cavendish, modern and ancient authors shared an epistemological problem: "They endeavour to deduce the knowledg of causes from their effects, and not effects from their causes, and think to find out Nature by Art, not Art by Nature."[36]

Cavendish conflated nature and reason. To reason naturally, and therefore understand something about the nature of nature, a philosopher should begin with rationality, not experimentation. Rationality itself consists of the material movements of what Cavendish calls "rational matter" and therefore creates what we might call first-order knowledge, rather than the second-order knowledge of the experimenters. Ironically, in her view, instead of denuding the personified Nature of her extraordinary properties, the new scientists create "praeternatural" and "artificial effects." Art or experimentation is "Natures foolish changeling Child," and the experimenters create "hermaphroditical" effects that are "partly Natural, and partly Artificial; Natural, because Art cannot produce anything without natural matter, nor without the assistance of natural motions, but artificial, because it works not after the way of natural productions."[37]

Rather than nature being full of unnatural, preternatural, or supernatural occult properties, Cavendish's matter includes vitality as part of its essence. Like Boyle and Descartes, Cavendish believed that the world consists of matter in motion, but unlike them, she thought moving matter was sentient, self-conscious, and self-moving. Instead of disenchanting nature, Cavendish wanted to unenchant it by incorporating soul and spirit into the material constituents of things. In her universe, minerals, vegetables, animals, and humans all possess, to some extent, sense and reason.

Some feminist critics would find Cavendish's embrace of vitalistic materialism only natural, at least in terms of current concepts of female psychology. These thinkers—most famously Mary Belenky and Evelyn Fox Keller—advocate "difference feminism," whereby women as a group are credited with a more holistic and empathetic understanding of nature; they have a particular "women's way of knowing."[38] But these feminists recognize that female attributes are culturally constituted, although they often fall into a universalist rhetoric. There was no necessary connection between sympathy for nature and female psychology in the seventeenth century, even when nature was depicted as a woman. As we saw above, Cavendish's use of gender categories, in both social and epistemological terms, was more complex and nuanced than any simple feminist interpretations would allow. When Cavendish integrated holistic ideas of matter into her philosophy and demonstrated an empathy with animals, she did so not because of any essentialist female epistemological imperative but because such ideas broadened the scope of nature to include all living beings.[39] Such a position allowed her both to present her own ideas and to critique the views of other natural philosophers.

Indeed, Cavendish had a profound sympathy for the natural world. She viewed the hunted animal with pity and the arrogance of man toward animals with disdain. "Man, out of self-love, and conceited pride," she writes, "thinks himself the chief of all Creatures, and that all the World is made for his sake; doth also imagine that all other Creatures are ignorant, dull, stupid, senseless, and irrational."[40]

Animals' materiality guarantees their rationality. Animals, like humans, are composed of both rational and inanimate matter; consequently "other Creatures have as much knowledg as Man. . . . But their knowledges being different, by reason of their different natures and figures, it causes an ignorance of each others knowledge."[41] Men, who cannot know the rationality of other creatures, conclude in their ignorance that other creatures have no reason and no soul. Some even think that animals are automatons who feel nothing, even when they suffer.

The debate over whether animals possess sense and soul is omnipresent in early modern literature—a complement to the *querelle des femmes* over whether women possess reason. Cavendish contributed to both discussions, and her reaction to the new science was conditioned by the broader discussion about the possibility of rationality in beasts and women. There are many references in her work to female "foolishness" or the "witless" nature of most women. Nature itself is sometimes blamed for their congenital deficiencies, but as we saw above, more often Cavendish points to the insufficiency of female education.[42]

Arguments about rationality are ultimately arguments about power. Nature—and women and animals—escape the confining categories of natural philosophers. Cavendish argued that when natural philosophers try to limit nature to certain principles, like semina or atoms, they reflect the paucity of their own conceptions and the smallness of their own minds, rather than true knowledge. "Atomists," because their brain or particular finite reason cannot reach further, "are much disceived in their arguments, and commit a fallacy in concluding the finiteness and limitation of Nature from the narrowness of their own conceptions." "Chymists," when they seek transformations in nature, are presumptuous and absurd: "Neither can there be any such thing as a new Creation in Nature."[43] Experimenters, when they dissect living bodies, cannot see the interior motions they claim to reveal. When they make knowledge claims that "the bare authority of an Experimental Philosopher is sufficient to them to decide all Controversies, & to pronounce the Truth without any appeal to Reason," they make a claim to power that both nature and reason defy.[44] "Modern Philosophers," in short, "make as great a noise to little purpose, as the dogs barking and howling at the Moon."[45]

Cavendish's wholesale repudiation of the practice, program, and claims of "modern philosophers" shows that the triumph of the scientific revolution was not uncontested. Certainly, the Royal Society was successful in establishing itself as the arbiter of knowledge. Cavendish's own contemporaries, including some women, viewed her with disdain and thought her insane. But in her condemnation of the new science, Cavendish spoke for many silenced voices, and it is within the context of that multitude that we must understand her. Cavendish spoke for nature against art/experimentation, for animals against humans, for the unlearned against the learned, for women against male arrogance and power.

To some extent, women and animals are always the "other" in any culture. Men are defined by what they are not—dog, ape, or bear—and human morality by what is considered unnatural or too natural.[46] Women fit uncomfortably into this binary division. If they act like men, they are hermaphrodites; if they act like animals, they are bestial. But their otherness gives them power. They are opaque, just as nature in its totality is hidden from humankind. "All Man-kind that ever have liv'd, or are at present living in this world," Cavendish insists, "could never find out the truth of Nature, even in the least of her parts."[47] When men think they understand nature, they are mistaken, and when they believe that their superiority to nature gives them dominance over it, they are wrong, "for it is a false Maxime to believe, that if some Creatures have power over others, they also have power over Nature."[48] If Nature could talk, she would say, "You cannot know me." If animals could talk, they would say, "What I know, you cannot know." If a female natural philosopher could be heard, she would say, "I know what Margaret Cavendish knows."

Cavendish represents the perspective of the stranger: someone who is outside a community but knows enough about it to understand it. Her aim was to undercut and challenge the claims of previous and current natural philosophers to preeminence and power. All of her many treatises on natural philosophy, her poetry and plays, and her science fiction represent a woman speaking to power or imposing her own power over subordinates.

Did Cavendish know that she was speaking for nature, animals, and women? Of course she did. Her tone is sometimes defensive or apologetic, but her work itself is never humble. Her many female avatars—the Empress of the Blazing World, the Lady Sanspareille, the She-Anchoret in a story of the same name— lecture men, and their speeches are not short. The man-beasts of the Blazing World are sometimes foolish but never at a loss for words. In fact, the capacity of animals to "commit mistakes and absurdities as Man" does testifies to their intelligence, if not their wisdom. "Wherefore," Cavendish concludes, "other Creatures may be Philosophers and subject to absurdities as aptly as Men."[49]

Margaret Cavendish fashioned herself as a philosopher, and so she had as much right as any to be absurd or to be profound. Foolishness is the human—and animal—condition, and, at least according to early modern conceptions of gender and rationality, women are the most foolish of all humans and animals. Perhaps that makes them particularly well qualified to understand nature and explain it to others. Hence, the ape-men and bear-men and lice-men, and the philosophers, applaud the knowledge of Cavendish's female projections. Facing a new learning that challenged tradition but kept her in her place, Cavendish confronted its authority and made her own scientific revolution.

# chapter one

# A Wonderful Natural Philosopher

*M*argaret Cavendish, Duchess of Newcastle, was a *stupor mundi*, a wonder of the world, a spectacle amazing and puzzling—then and now. When she went to London to visit the Royal Society, Samuel Pepys described "100 boys and girls running looking upon her," while at the same time he desperately attempted to view her himself. "There is as much expectation of her coming to Court, that so many people may come to see her," he writes, "as if it were the Queen of Sweden."[1] When he attempted to catch sight of her in Hyde Park, he complained, "That which we, and almost all went for, was to see my Lady Newcastle; which we could not, she being followed and crowded upon by coaches all the way she went, that nobody could come near her; only I could see she was in a large black coach, adorned with silver instead of gold, and so with the curtains, and everything black and white and herself in a cap."[2]

Cavendish was striking in appearance and striking in action. Women did not write natural philosophy in the seventeenth century. To do so was not only revolutionary but even unnatural, a complete blurring of the gendered characteristics considered inherent in the male and female. Yet Margaret Cavendish wrote, among other works, two books of scientific poetry and five treatises about natural philosophical questions. She developed a unique natural philosophy based on the assumption that the world was composed of animated and internally self-conscious matter in motion. Her vitalistic materialism contained echoes of the other natural philosophies of the early seventeenth century, but these ideas were always filtered through the prism of Cavendish's imagination and reflected seventeenth-century gendered assumptions. Part of Cavendish's role was the reception, reconstitution, and dissemination of the ideas of her male counterparts. Part of her own self-fashioning was integrating elite and popular notions of women into her role as a writing woman. Part of her role as a female natural philosopher was to emphasize a more organic and sympathetic view of nature, which she characterized as a powerful and fecund woman or, alternatively, as the internalized ordering principle of matter. Her natural philosophic stance brought her into direct conflict with the emerging science of the Royal Society.[3]

In Cavendish's forays into philosophy, both intellectual and personal, she made a virtue out of necessity—substantiating her role as a natural philosopher by embracing an epistemology that validated any judgment, whether male or female, in philosophic reasoning and by arguing for the finer perception of female imaginative faculties. Such a conclusion about female ability, however, was not adopted without a certain degree of uncertainty and self-doubt. Cavendish understood that in becoming a female natural philosopher she was becoming a kind of monster, facing the monster's fate of being displayed and despised in the circus of the world.[4] Ultimately, she responded to this threat by attempting to subvert the male fraternity of the Royal Society.

In a classic article of second-wave feminism, Joan Kelly argued that "every learned tradition was subject to feminist critique, since all were dominated by men and justified male subjection of women."[5] We will return to the question of Cavendish's feminism in the last chapter of this book, but here it is enough to recognize that a "female" interpretation of natural philosophy is not necessarily an expression of power or its subversion—although ultimately a "new" philosophy can lead to a new understanding of the place and potential of women. Modern critics who emphasize Cavendish's radicalism and feminism, and those who deny her these stances, fail to recognize the complexity of her work and her position in the social and intellectual life of mid-seventeenth-century England.[6]

Diverse elements go into the making of a thinker. As discussed in the Introduction, Cavendish knew the ideas of other philosophers and scientists through discussion with her family. In the 1660s, she embarked on an active reading of philosophic and scientific works, including writings of Descartes, Hobbes, and Boyle.[7] But Cavendish was not a passive vehicle for the expression of others' thoughts; she actively recast what she learned and fashioned her own natural philosophy. She was inspired by the assault on the traditional learning of the schools, but her own position as a royalist member of the upper classes made her cautious about treating any traditional authority with disdain. Issues of class as well as gender complicated her approach to natural philosophy.

The attack on authority in the late sixteenth century, and the formulation of new philosophies in the early seventeenth century, empowered Cavendish to question traditional philosophic certainties and traditional female roles. The underlying epistemological premise of Cavendish's natural philosophy is a form of extreme skepticism about the possibility of absolute knowledge of nature.[8]

Cavendish felt that our senses cannot penetrate nature's mysteries, and since every age believes something different, "But all Opinions are by Fancy fed,/And Truth under opinions lieth dead."[9] Cavendish, at least in her early works, felt that there was a continuum between imagination, or what she calls fancy, and reason and that women were particularly suited to explore the connection. Thought, for example, could be explained by the actions of fairies in the brain: "Who knowes, but in the Braine may dwel/Little small Fairies; who can tell?"[10]

A skeptical epistemology was particularly congenial to a woman who wanted to write natural philosophy. It was liberating for Cavendish, licensing her methodology of "sense and reason," or wide-ranging speculation based on unmediated observation combined with rational abstraction and imaginative speculation. Her work shows that the new philosophy had radical implications far beyond the purview of science. It became a tool of self-fashioning for a woman trying to define a new construct: a woman natural philosopher.

Cavendish's work, like herself, combined elements usually distinct in the seventeenth century. The dual expression of Cavendish's ideas in philosophic and imaginative works reflected a self-conscious effort to reach a broader audience than any male learned thinker had either contemplated or desired. Cavendish was truly a cultural hermaphrodite whose liminal position between elite and popular culture allowed her to transverse and unite the worlds of the most educated, the semieducated, and the least educated.

Cavendish was not only a writing woman, she was also a natural philosopher and a fabulist; in her works, she was humorous and she was serious, she was practical and she was fanciful, she wrote essays and romances, poetry and treatises. But rarely has an author been more self-conscious—or more uneasy—about the ambiguity of her own work and her own role. She repeatedly tells the readers it is their responsibility to make sense out of her concoctions, not hers. For example, in *Natures Pictures*, she proclaimed, "Also I am to let my Readers to understand, that though my work is of Comicall, Tragicall, Poeticall, Philosophicall, Romanicicall, Historicall and Morall Discourses, yet I could not place them so exactly into severall Books, or parts as I would, but am forced to mix them one amongst another, but my Readers will find them in the volume, if they please to take notice of them, if not there is no harme done to my Booke, nor me the Authoress."[11]

Cavendish knew it was odd to mix genres, but she justified it on the grounds that fancy "recreate[s] the Mind." Both reason and imagination characterize the rational part of matter—matter itself becomes defined by its epistemological properties. Thus, the forms of cognition subsumed within rational matter include reason, "a rational search and enquiry into the causes of natural effects," and fancy or imagination, "a voluntary creation or production of the Mind." In a sense, Cavendish materialized the faculties of intellect and will and then transformed them into a stylistic aesthetic. Hence, she explained why she published her critique of experimentalism, *Observations upon Experimental Philosophy* (1666), in the same volume as a romantic fantasy, *The Blazing World*: "This is the reason, why I added this Piece of Fancy to my Philosophical Observations, and joined them as two Worlds at the ends of their Poles; both for my own sake, to divert my studious thoughts, which I employed in the Contemplation thereof, and to delight the Reader with variety, which is always pleasing."[12]

Paradox is an essential quality of Cavendish's works, reflecting not only her skepticism but also her ambivalence about categories of literature and categories of gender.[13] All of Cavendish's works reveal a basic structural similarity, in which reason and imagination function as dual expositors of her philosophy. The different sections of *Poems, and Fancies* (1653) are literally pinned together by what Cavendish called claspes; they are meant to be read as a unit, with the fanciful parts commenting and expanding on the more philosophic sections. Likewise, *Philosophicall Fancies* (1653) begins with a discussion of Cavendish's developing vitalistic materialism and concludes with imaginative poems exploring the likelihood of different kinds of animate beings, including flower women and metal men. *Philosophical and Physical Opinions* (1655), which contains a detailed exploration of Cavendish's new natural philosophy, should be read in conjunction with *Natures Pictures* (1656), a collection of tales and romances that projects Cavendish's thought to a more popular audience and includes accounts of imaginary journeys that clearly anticipate the imaginative universe of *Blazing World*. *Philosophical and Physical Opinions* was reprinted in a vastly rewritten text in 1663; several of its themes are treated fictionally in the 1662 edition of her *Plays*. And in 1664, Cavendish published her *Sociable Letters* and *Philosophical Letters*; the first included an exploration of social, political, moral, and cultural themes, while the second was a critique of other recent natural philosophies.

Cavendish knew that in writing these works she was flouting the social and cultural expectations of her society; she often reflected on the propriety of her own actions as a writing woman and on the possibility of female intelligence. In 1664, she wrote, "It may be said to me, as one said to a Lady, Work Lady, Work, let writing Books alone . . . the truth is, My Lord, I cannot Work, I mean such Work as Ladies use to pass their Time withall."[14] In *Poems, and Fancies*, she defended herself from the claim of being really odd: "Yet I have Truth to speak in my behalfe for some favour; which saith first, that Women writing seldome, makes it seem strange, and what is unusuall, seemes Fantasticall, and what is Fantasticall, seemes odd, and what seemes odd, Ridiculous: But as Truth tells you, all is not Gold that glisters; so she tells you, all is not Poore, that hath not Golden Cloaths on, nor mad, which is out of Fashion."[15]

In combining and using different genres to express her ideas about nature, Cavendish contested the new rhetorical orthodoxy of the Royal Society. Francis Bacon had cautioned his followers against allowing poetry, romances, fables, "and frivolous impostures for pleasure and strangeness" to corrupt "a substantial and severe collections of the heteroclites and irregulars of nature." Bacon believed that each kind of learning had its own particular form of expression, and although poetry might be perfectly congenial to romance, it was inappropriate for natural history.[16] Such genre limitations were inimical to Cavendish's style and philosophy; her profuse style also contributed to her exclusion from the new forms of scientific expression.

Some of Cavendish's readers suspected that her rhetorical style was affected by scholastic conceits, not a commendable practice in the mid-seventeenth century. After the publication of *The Worlds Olio* (1655), the physician Walter Charleton informed Cavendish that some readers doubted her originality, "And the Reason they give me, is this; that you frequently use many Terms of the Schools, and sometimes seem to have Imp'd the Wings of your high-flying Phancy with sundry Feathers taken out of the Universities, or Nests of Divines, Philosophers, Phisicians, Geometricians, Astronomers, and the rest of the Gowned Tribe."[17]

Newcastle also defended his wife from the charges of irrationality and imitation. At the beginning of the second edition of *Philosophical and Physical Opinions* (1663), he argued that philosophers,

Their Judgments, Understandings, Drowned so,
Because they Writ of what they did not Know,
Were all but Guessers, then this Lady may
Write her Opinions, Better than all they,
And Set up for her Self, for all may find,
Reading her Book, 'twill make them of her Mind,
So Rational, such Solid Judgments hight,
As all Wise Men will Swear, that it is Right.[18]

Newcastle repeated the claim that Cavendish learned the terms of her discourse from his brother, Sir Charles, and him. He contended, however, in the 1655 *Philosophical and Physical Opinions* that Cavendish could easily have picked up the language of her philosophy from the world around her. Theological terms are so common that "every Tub-preacher discourses of them, and every sanctified wife gossips them in wafers, and hipocris at every Christening." Likewise, medical vocabulary is known by any "good Farmers wife in the country," geometrical terms are known "even to Joyners and Carpenters," and astronomical language is easy to acquire, "since every boy may be taught them, with an apple for the Globe, and the parings for the sphears." Thus, "it is so ridiculous then to think that this Lady cannot understand these tearms, as it is rather to be laught at, then to trouble ones self to answer."[19]

According to the (soon-to-be) duke, philosophic terminology, if not philosophic understanding, had permeated popular culture. Newcastle might have shied away from the implication that his wife shared the culture of farm wives, artisans, and boys, but such a view also allowed him to dismiss the intellectual exclusivity of the learned and endorse his wife's philosophizing. Such an attitude underscored the subordinate position of scholars, women, and the common man, none of whom were considered gentlemen—although Steven Shapin and Simon Schaffer have shown that the scholars, at least, were trying to improve their social standing.[20]

As a conservative royalist, the duke recognized that the permeability of intellectual boundaries in the mid-seventeenth century could be subversive. In a letter of advice he wrote Charles II shortly before the Restoration, Newcastle counseled the king to use the recreations of the "meaner People" to increase his power and distract them from potentially dangerous discourse: "The devirtismentes will amuse the peoples thaughts, & keepe them In harmless action which will free your Majestie frome faction & Rebellion." He listed the pabulum of the people: "Butt five or Six Playe Houses Is Enough for all sortes off Peoples divertion & pleasure In thatt kinde,—Then Pupett playes Ther will bee to please them besides,—As also dansers off the Ropes with Guglers & Tumblers,—Besides strange Sightes, off Beastes, Birdes, Monsters, & manye other thinges with severall Sortes of Musike & dansinge,—Ande all the olde Holedayes with their Mirth, & rightes sett upp agen; Feastinge dayleye will bee in Merrye Englande."[21]

The Duke of Newcastle knew that the boundaries of culture in "Merrye Englande" had to be policed; when peasants had become theologians and astronomers, the result had been civil war. Keeping peasants in their place was crucial, but so was containing the presumption of experimental philosophers who claimed that their games revealed at least the probable truths of nature. The pretensions of both natural philosophers and peasants could be controlled by diverting speculation into entertainment. Cavendish wrote of her husband, "Nor doth he think it a crime to entertain what Opinion seems most probable to him, in things indifferent; for in such cases men may discourse and argue as they please, to exercise their

Wit, and may change and alter their Opinions upon more probable Grounds and Reasons."[22] Newcastle reflected a dominant view of natural philosophy in the late Renaissance, when, according to Paula Findlen, "science . . . was a divinely inspired guessing game in which natural philosophers attempted to infer what neither God nor nature would ever tell them." It was a kind of game or play that investigated the wonders and diversity of nature—the jokes of nature (*lusus naturae*)—in order to reveal God's or Nature's extraordinary fecundity.[23] In *Observations*, Cavendish wrote, "But wise Nature taking delight in variety, [so] her parts, which are her Creatures, must of necessity do so too."[24] Just as Cavendish herself enjoyed different genres, Nature delighted in her many different creations.

Thus, the study of nature becomes a kind of guessing game, as Newcastle proclaimed in the preface to the 1663 edition of Cavendish's *Philosophical and Physical Opinions*: "Since now it is A-la-Mode to Write of Natural Philosophy, and I know, no body knows what is the Cause of Any thing, and since they are all Guessers, not Knowing, it gives every Man room to think what he likes, and so I mean to Set up for myself, and Play at this Philosophical Game as follows, without Patching or Stealing from any Body."[25]

The wide scope of science as a game allowed Newcastle to destroy the pretensions of natural philosophers and endorse the purpose of the study of nature as recreation. He did not seem to worry that such devaluing of natural philosophy implicitly undermined his wife's writings. It is not surprising that in one of the dedications—"To Natural Philosophers"—in her first book, *Poems, and Fancies*, Cavendish fashions her enterprise as an entertainment:

> I desire all that are not quick in apprehending, or will not trouble themselves with such small things as Atomes, to skip this part of my Book, and view the other, for feare these may seem tedious: yet the Subject is light, and the Chapters short. Perchance the other may please better; if not the second, the third; if not the third, the fourth; if not the fourth, the fifth: and if they cannot please, for lack of Wit, they may please in Variety, for most Palates are greedy after Change. And though they are not of the choicest Meates, yet there is none dangerous; neither is there so much of particular Meat, as any can feare a Surfet; but the better pleas'd you are, the better Welcome. I wish heartily my Braine had been Richer, to make you a fine Entertainment.[26]

Figuring her scientific poetry as a banquet prepared by a woman for the delight of natural philosophers made Cavendish the agent and other thinkers the participants in a festival of fictive and abundant creativity. To Cavendish, such an enterprise is a "harmlessest pastime," unlike the actions of some women during the civil war, who entwine themselves into the "Politicks of State, or to Preach false Doctrine in a Tub."[27]

As a form of recreation, natural philosophy shared the cultural function of other kinds of entertainment. All presented wonders, secrets, and fabulous monsters. Cavendish enjoyed the spectacle and gender-bending permitted during times of carnival and holiday. In a letter written while she lived in exile in Antwerp, she described carnival as "the most Pleasant and Merry time in all the Year." Masking, in particular, delighted her: "But truly, these are Harmless Sports, consisting only in several Attirres, or Accoustrements, as to wear Vizards, &c. and some of the Women do Accoustre themselves in Mens Habits, and the Younger sort of Men in Womens Habits, where the Women seem to be well Pleased, and take a Pride to be Accoustred like Men, but the Men seem to be more out of Countenance to be Accoustred like Women, as counting it a Disgrace to their Manhood."[28] Cavendish's sense of irony did not inhibit her enjoyment, although she did claim that "several Sights and Shews" of the city, like "Dancers on the Ropes" and "Acting Baboons, and Apes," "which would be as Tedious to me to Relate as to See, for I would not take the pains to See them."[29]

Cavendish, like her husband, knew that carnivals and fairs could work to reaffirm traditional social hierarchies. After a period of play, the participants could return to their conventional roles. The historian Natalie Davis has argued that in the rigid society of early modern Europe festivals were temporary respites that allowed people to blow off steam and reaffirmed conventional roles.[30] Other historians, however, have recognized the potential subversiveness of such occasions, when some people acted out their challenges to cultural norms. A game could turn into a riot. Natural philosophers playing with nature could become rebels against traditional authority. And a female natural philosopher could use play and spectacle to subvert old institutions and new. By creating display and putting herself on display in visiting the Royal Society, Cavendish achieved three potentially paradoxical aims: she showed that experimental philosophy was an inconsequential game, that the practitioners of experimental philosophy were déclassé, and that a woman had as much right to practice natural philosophy as any man.

## AN UNCONVENTIONAL HOUSEWIFE

Cavendish used both her writing and her person to evoke astonishment and even repugnance in her audience by composing scientific poems and long philosophic treatises and by embracing eccentricities of dress and deportment. Her contemporary Dorothy Osborne commented on the publication of *Poems, and Fancies,*

"For God's sake, if you meet with it, send it me; they say 'tis ten times more ex-travagant than her dress. Sure, the poor woman is a little distracted, she could never be so ridiculous else as to venture at writing books, and in verse too."[31]

In publishing her own opinions and fancies, Cavendish turned cultural cate-gories upside down.[32] Dorothy Osborne thought her mad as well as ridiculous. Many others since have shared this opinion.[33] But Cavendish chose to make her eccentricity an asset. The hierarchy she critiqued and challenged in her works was what A. O. Lovejoy famously called "The Great Chain of Being."[34] This con-ceptualization derived from the Greeks and integrated the cultural presupposi-tions of both the elite and nonelite classes, locating all being along an ascending (or descending) chain of worth. The nonsentient, rocks and minerals, were at the bottom, while the most self-aware, God and the angels, were at the top. Every-thing else fell between, with man the link between mortal and immortal. Women traditionally were viewed as the link between man and the animals, human be-ings but closer to the earth and nature.[35]

Cavendish understood her role of mediator between man and nature and self-consciously sought an audience for her works beyond the educated elite. "The truth is, if any one intends to write Philosophy, either in English, or any other language; he ought to consider the propriety of the language, as much as the Subject he writes of; or else what purpose would it be to write?" She, unlike other natural philosophers, has considered the problem of intelligibility: "Although I do understand some of their hard expressions now, yet I shun them as much in my writings as it is possible for me to do, and all this, that they may be the better understood by all, learned as well as unlearned; by those that are professed Phi-losophers as well as by those that are none."[36] While many of Cavendish's fellow philosophers professed a similar rhetorical simplicity, their aim was to distance themselves from the obscurity and complexities of scholastic writing. Cavendish's goal, however, was to make her philosophy available to a much broader public who, like her, would come to understand the new ideas. Thus, she explicitly and paradoxically challenged the gender and class hierarchies she and her husband elsewhere affirmed.

The execution of this plan of cultural dissemination was somewhat under-mined by her prolix style and creative terminology. It is difficult to accept the notion that Margaret Cavendish believed that the unlearned would pick up one of her dense philosophic treatises and absorb it. But possibly she thought her more fantastical works would be accessible to those who could read, or at least to those who could afford expensive books, even if they were not scholars and professors. In her *Orations of Divers Sorts*, a series of imagined speeches on government and society, one of Cavendish's characters, a peasant, argues that farmers are akin to Nature in that both can create and both can commit errors, and the human mistakes are made because "though we are Bred up to Husbandry, yet we are not all so Knowing in Husbandry, as to Thrive and Grow Rich by our Labours." In fact, the peasants' lack of knowledge is like the ignorance of scholars "that have Lived and spent most of their time in Studies in Universities, but are meer Dunces." Peasants in fact can overcome their deficiency by bringing together "Practice and Wit," and "Thus Clowns, Boors, or Peasants by Name, are become Princes in Power."[37]

Both peasants and women can overcome their challenges by combining their wit and their experience, perhaps more successfully than scholars who remain dunces. Cavendish was cognizant and proud of her connection with the more "natural" perceptions of the untutored, and she believed that nature itself validated her use of imagination to understand nature. True, she acknowledges, she has deviated from her expected gender and social role: "The truth is, I have somewhat Err'd from good Huswifry, to write Nature's Philosophy, where had I been prudent, I should have Translated Natural Philosophy into good Huswifry."[38] She knew that women's "work" is "Needle-works, Spinning-works, Preserving-works, as also Baking, and Cooking-works, as making Cakes, Pyes, Puddings, and the like, all which I am Ignorant of."[39] In some of her works, Cavendish implicitly resolved the contradiction between what she was and what she should be by making a female Nature the mediator between creation and God: "For Nature, being a wise and provident Lady, governs her parts very wisely, methodically and orderly; also she is very industrious and hates to be idle, which makes her imploy her time as a good Huswiffe doth, in Brewing, Baking, Churning, Spinning, Sowing &c. as also in Preserving . . . and in Distilling . . . ; for she has numerous imployments, and being infinitely self-moving, never wants work, but her artificial works are her works of delight, pleasure, and pastime."[40]

By imagining Nature as a "provident Lady," Cavendish integrated cultural norms into the basic fabric of her natural philosophy, not only finding a way to justify her activity but also shaping the very content of her material philosophy. Instead of being a housewife of the home, Cavendish was a housewife of ideas; for her not to fill this role would be as unnatural as nature not governing the world. Thus, Cavendish refigured the popular image of woman as housewife to justify her own metamorphosis from woman to writer. At the same time, she sought to explain the deficiencies she saw in other women as the result of men seeking to make them into monsters or beasts or peasants by denying them education. Educated women were anomalies, and Cavendish knew that her efforts left her open to charges of being not only singular but even monstrous.

## THE VISIT TO THE ROYAL SOCIETY REVISITED

Intellectual women were wonders or prodigies in early modern times. According to Margaret L. King, "like divine miracles, they were both wondrous and terrible; as prodigies, they had exceeded—and violated—nature. Male by intellect, female in body and in soul, their sexual identity was rendered ambiguous. . . . Not quite male, not quite female, learned women belonged to a third and amorphous sex."[41]

Just as monstrosities were supposed to evoke wonder—either pleasure or horror—a writing woman could expect only astonishment and ridicule.[42] She was by definition unnatural, a kind of hermaphrodite who flouted the cultural expectations of society, whether low or high. Cavendish could not escape this *mentalité*. It colored the way she viewed herself and women in general. It expressed itself in anxiety and defensiveness about the possibility of female intelligence and her own role as a writing woman. Ultimately, however, Cavendish turned her deviancy into an advantage. By exhibiting herself as a spectacle in visiting the Royal Society, she diverted the public gaze from the sober scientists of the society, implicating that they were as much mountebanks as the hucksters Newcastle thought should entertain the masses.

Cavendish never explicitly stated her objective in attending a meeting of the Royal Society. Most commentators have interpreted her actions during the visit as an example of a woman overwhelmed by what she observed, rendered literally almost speechless—except for saying, according to Pepys, that she was "full of admiration, all admiration."[43] In casting Cavendish as the passive observer, such interpretations ignore both the attacks on the society she published before and after the visit and the fact that her intended audience for the outing was not only the experimenters but society at large. Her attendance at the Royal Society necessitates a historical semiology that interprets acts as well as words.

Cavendish's very public visit must be seen within the context of the performative culture to which she belonged. According to Rebecca D'Monté, Cavendish is emblematic of "the idea of staging the self or recreating oneself through performance . . . that was endemic to seventeenth-century culture, a period when bodies 'made a display of themselves.'"[44] Display in the seventeenth century could operate on a number of levels. There were the moments of ritualized display during carnival, when social categories were overturned; there were fairs and circuses where monstrosities and oddities were presented to amazed crowds; and there were the displays catering to the upper classes, like the museums of curiosities collected by nobles and gentlemen. There were also the pageants and parades used by kings and lord mayors to present their sanctified selves to the gaze of the adoring public. In Antwerp, where Cavendish lived between 1649 and 1660, she sometimes went on the "Tour," a circuit of the town "where all the chief of the Town goe to see and be seen, likewise all strangers of what quality soever."[45]

Cavendish had access to all these arenas of spectacle. Peter Burke has suggested that "perhaps one could see noblewomen as mediators between the group to which they belonged socially, the elite, and the group to which they belonged culturally, the non-elite."[46] While educated men pursued the learning of the ancients and moderns and, as Burke has argued, participated in the rituals and activities of the lower classes as a form of play, the semieducated women of the upper classes were forced into a level of synergy with "the people" that affected the way they knew and understood the world.[47] Cultural historians no longer accept such a reductionist model, but clearly upper-class women had a closer cultural, if not sexual, relationship with the lower classes than did their husbands, fathers, and brothers. In her autobiography, Cavendish recounts how her mother tried to keep her daughters from conversing with nurses and servants, recalling her mother's belief that "the vulgar sort of servants, are as ill bred as meanly born, giving children ill examples, and worse counsel."[48] Nevertheless, nurses and servants are an active and vital presence in her plays. Thus, as an informally educated upper-class lady, Cavendish brought the assumptions and presuppositions of popular culture to her reading and writing of natural philosophy.[49] Her sex allowed her to play with the conventions of popular and elite culture, and especially to subvert gender expectations in her work and behavior.

Cavendish acted as if the gender-bending she observed during carnival and fairs was appropriate even when she too was the object of spectacle. The diarist John Evelyn's description of Cavendish's visit to the society captures her extraordinary masculinized appearance:

Her head-gear was so pretty
I ne'ere saw anything so witty
Tho I was halfe a feard
God blesse us when
I first did see her
She look'd so like a cavaliere
But that she had no beard.[50]

Cavendish's dress incorporated elements of male clothing, problematizing her gender and potentially making her a monster, like the carnival oddities she had viewed in another cultural space. Beardless men were at best eunuchs, at worst hermaphrodites who were often included as roadside attractions in public celebrations. In *Orations* (1662), Cavendish included a debate during which several women discuss the position of women. After one woman suggests, "We should imitate men, so will our bodies and minds appear more masculine, and our power will increase by our actions," another replies, "To have female bodies, and yet to act masculine parts, will be very preposterous and unnatural; in truth, we shall make ourselves like as the defects of nature, as to be hermaphroditical, as neither to be perfect women nor perfect men, but corrupt and imperfect creatures."[51]

Cavendish clearly understood that in adopting male dress and male activities she made herself vulnerable to condemnation.[52] In the 1655 edition of *Philosophical and Physical Opinions*, Cavendish argued that if she did not defend the originality of her natural philosophy, "it may be thought I were not a right begotten daughter of nature, but a monster produced by her escapes, or defects; for every true childe of nature will require its just inheritance."[53] Cavendish's uneasiness is clear: nature does produce monsters; every country fair had its sports and wonders, but she is not one of them.

As we have seen above, the structure of Cavendish's work was hermaphroditical. Many commentators have remarked on the prevalence of such images in her writing. Her plays and stories are replete with cross-dressing and ambiguous females; her anxiety about her unnatural action in writing produces constant apologia to her readers, to whom she sought to explain and justify herself.[54] But her work also embraced the possibility that exceptional women can win respect and power, that they are admirable. By transgressing gender borders, they are granted public triumphs, become empresses, and are worshiped like gods.[55] They become wonderful as well as monstrous.

Wonder was a complex emotion in early modern Europe. According to Lorraine Daston and Katharine Park, wonders evoked both pleasure and fear in spectators. They were a source of delight and amazement: "To register wonder was to register a breached boundary, a classification subverted. . . . The domain of wonder was broad, and its contexts were as various as the annual fair, the nave of a cathedral, the princely banquet hall, the philosopher's study, or the contemplative's cell." As elite and popular culture became more distinct in early modern Europe, members of the educated classes increasingly viewed wonders with "impassivity or downright repugnance."[56] Cavendish could easily accommodate her experience of novelty and strangeness with pleasure, but the members of the Royal

Society could view spectacle only with repugnance as they tried to create a new identity as gentlemen-scholars and the only legitimate interpreters of nature.[57]

In the sixteenth and early seventeenth centuries, before the institutionalization of science, one of the most important aspects of the study of nature was its capacity to astonish and thereby entertain.[58] We have seen that William Newcastle viewed science as a parlor game; Margaret Cavendish saw experimental philosophy as a kind of play and display, both figuratively and literally. Thus she referred to experimenters and natural philosophers as boys playing with toys.[59] Nothing could have been more antithetical to the dogged seriousness of the Royal Society, but natural philosophy as entertainment provided an entry point for the participation of a writing woman in the scientific enterprise. By visiting the Royal Society in all her cross-dressing splendor, Cavendish could reinforce her claim to be the equal or even superior of her fellow investigators of nature. By emphasizing the performative nature of experimental science, Cavendish turned the role of scientist upside down: the scientists' theater became her stage. The experimenters became sideshow entertainers, equivalent to the fake doctor Cavendish had seen at a fair. In *Sociable Letters* she writes, "But where there is one Feigned Fool in the World, there are a thousand more Feigned Wise men, and where there is one Professed mountebank, or Jugler, there are thousands more that are so, but will not be Known or Thought to be so."[60] She could have been writing about the Royal Society.

The members of the society knew that they were obliged to entertain, but their idea of speculative wonder repudiated both the sideshow entertainer and the aristocratic dilettantes who played with nature. As Shapin and Schaffer have shown, the "witnessing" of experiments in the public space of the society was essential to the scientific program of the new experimentalists.[61] Only through the collective speculative gaze of the society could knowledge become legitimate. The purpose was not entertainment, although the observation of the ordinary course of nature, rather than the extraordinary, did produce delight. Although "entertainments" could be produced to amuse visitors to the society, like Cavendish and Charles II, they should have serious import. "For," wrote Robert Boyle, displaying some amount of social—and theological—anxiety, "the works of God are not like the Tricks of Juglers or the Pageants that entertain Princes."[62] Instead, Boyle explained, experiments served only to gain the interest of those observers who might then become interested in "much more difficult" studies like "Chymistry and Corpuscular Philosophy." In his *Experiments and Considerations Touching Colours* (1664), a text Cavendish had read, Boyle proclaimed that even "Ladies" and "such kind of persons, as value a pretty Trick more than a true Notion, and would scarce admit Philosophy, if it approach'd them in another Dress; without the strangeness or endearments of pleasantness to recommend it." These kinds of virtuosos can be intrigued by "easie and recreative Experiments." Serious experimenters like Boyle, of course, "love to measure Physical things by their use, not their strangeness, or prettiness."[63]

Thomas Sprat, in his *History of the Royal Society*, written in the same year Cavendish visited the society, distinguished the aesthetic experience of observing the ordinary in nature from the transient pleasure of observing the "greatest Curiosities":

> For it [curiosities] will make men inclinable to bend the Truth much awry, to raise a specious Observation out of it. It stops the severe progress of Inquiry: Infecting the mind, and making it adverse from the true Natural Philosophy: It is like Romances, in respect to True History; which, by multiplying varieties of extraordinary Events, and surprizing circumstances, makes that seem dull and tasteless. And . . . the very delight which it raises, is nothing so solid: but as the satisfaction of Fancy, it affects us a little, in the beginning, but soon wearies and surfeits: whereas a just History of Nature, like the pleasure of Reason, would not be, perhaps so quick and violent, but of farr longer continuance, in its contentment.[64]

Monsters, romances, fancy are all condemned as antithetical to genuine natural philosophy. The passionate gaze of wonder has been displaced by the dispassionate analysis of matters of fact, even on the many occasions when the society did in fact consider "the most unusual and monstrous forces, and motions of matter." Sprat claimed that such phenomena are "admirable," that is, they are worth looking at both "to be fully acquainted with their compositions and operations" and to recognize those things which are truly extraordinary." However, natural history should not be like a romance: "To make that only consist of strange, and delightful Tales, is to render it nothing else but vain and, and, and ridiculous Knight-Errantry."[65] Cavendish, a year before Sprat's book appeared, had published her attack on the experimentalism of the Royal Society, with the romance *The Blazing World* appended to it. Her person, her work, her aesthetic were all equally repugnant to the practitioners of the new science.

Samuel Pepys, in his first mention of Margaret Cavendish, wrote, "The whole story of this Lady is a romance, and all she doth is romantic."[66] This compliment was a double-edged one, as was the reaction of John Evelyn, who wrote in his *Diary*, "I was much pleased with the extraordinary fanciful habit, garb, & discours of the Duchess." His wife, Mary, had a less positive impression: "Her mien surpasses the imagination of poets, or the description of the romance heroines greatness. . . . My part was not to speak, but to admire."[67] Fascinated by this extraordinary woman, "very much company" gathered at the Royal Society, recounts Pepys, "in expectation of the Duchesse of Newcastle, who had desired to be invited to the Society, and was, after much debate pro and con, it seems many being against it, and we do believe the town will be full of ballets of it."[68]According to Frances Harris, Mary Evelyn's letter describing her surprise "to find so much extravagancy and vanity in any person not confined within four walls" was intended to be passed around the Oxford colleges; it "was intended as a rebuke to all these men of science and learning for making any acknowledgement of her pretensions."[69] Clearly some members of the Royal Society—or their wives—feared that a visit by the notorious Margaret Cavendish would bring ridicule on an institution that was trying to establish itself as a serious center of study, rather than a place where some came "only as to a Play to amuse themselves for an hour or so."[70]

Cavendish's notoriety preceded her. A spectator wrote his father, "The Duchess of Newcastle is all the pageant now discoursed on: Her breasts are all laid out to view in a play house with scarlet trimmed nipples."[71] Her movements were meant to mimic the processions of royalty; indeed, the Lord Chamberlain forbade her liveries wearing affected velvet caps like the king's footmen. They were meant to impress the spectators, both the common people who thronged her and the courtiers who jockeyed to see her in Hyde Park. No wonder a hundred boys and girls ran after her.

Cavendish knew her visit to the Royal Society would cause a spectacle. "It is most certain," she wrote in the dedication to her 1668 edition of *Plays*, "that those that perform Public Actions, expose themselves to Publick Censures." It was worth the censure, which she already knew she had earned from some through the publication of her books, if it would redound on the society that had acquiesced to her desire to attend a meeting. Here is Pepys's description of the visit: "Anon comes the Duchesse, with her women attending her; among others, that Ferrabosco of whom so much talk is, that her lady would bid her show her face and kill the gallants. . . . The Duchesse hath been a good comely woman; but her dress so antic and her deportment so unordinary, that I do not like her at all, nor did I hear her say anything that was worth hearing, but that she was full of admiration, all admiration."[72] To this account, John Evelyn added that "she came in great pomp."[73] Cavendish's unordinary behavior and antic dress were both calculated to cause a sensation, a sense of wonder and amazement that would preempt and minimize the displays of the Royal Society and consequently diminish its program and promote her own.[74] Cavendish essentially had turned the meeting of the Royal Society into a carnival, with the exhibit being not the sober experiments with which she was entertained but herself at her most fantastic.

Cavendish's visit to the society took place on May 30, 1667. She had indicated earlier in the month that she wanted to be invited, and on May 23 Lord Berkeley and an old friend of the duchess, Dr. Walter Charleton, were instructed to inform her that she would be welcome at the next meeting of the society. Robert Boyle was asked to prepare an "entertainment" for her, including the use of the air-pump and various chemical experiments.[75]

The duchess kept the society waiting while a huge crowd gathered to see her enter. She was conducted to the meeting room by the president of the society, Lord Brouncker, who carried the royal mace, which had been presented by Charles II, before her. The visit had become a pageant. She was seated next to the president, who doffed his cap, although the provisions of the society allowed him to keep it on.[76] It was clear that the society was attending her, rather than she attending the society. As she watched, a young child ran up and down the room laughing.[77] He knew that he was attending a show, even if the Royal Society did not.

While Cavendish watched, Boyle expelled air from an air-pump, which was then weighed, and he demonstrated how water would bubble away in an empty-ing globe and how a bladder would inflate under the same conditions. He mixed colors to produce a third color, and he showed her the action of a terrella, or earth-shaped magnet.[78] For her delight, he dissolved roasted mutton with sulfuric acid, which was then transformed into, as Pepys says, "pure blood—which was very rare."[79] Finally, she was shown the use of "a very good microscope" and peered at the magnified body of a louse.[80] That was as close to animal experimen-tation as Cavendish got. Although in the meetings before and after her visit ex-periments on dogs and cats were discussed and performed, Cavendish was not privileged to see any. Perhaps some members of the Royal Society had read Cav-endish's *Poems, and Fancies*, in which, in her most famous poem, she expressed her sympathy for the hare pursued by hunters.[81]

According to Pepys, the only thing Cavendish had to say in response to these rare displays was that she was "full of Admiration, all Admiration." *Admiration* was an equivocal term in the seventeenth century. It could certainly mean "reverence, esteem, approbation," but it could also simply denote "wonder" and "astonish-ment."[82] As we saw above, Sprat decreed that the experimenter should be very cautious about the object of admiration. Boyle was likewise cautious with the use of "wonder" to elicit "admiration." In *Experiments*, he describes showing "trifles" to "many persons of differing Conditions, and ev'n Sexes," which filled the observers with "Admiration." "But," he adds, "'tis fitter for Mountebancks than Naturalis to desire to have their discoverys rather admir'd than understood, and for my part I had much rather deserve the thanks of the Ingenious, than enjoy the Applause of the Ignorant."[83]

Boyle's dismissal of "ladies" may have inspired Cavendish's plan to ridicule the society by associating its program with the tricks presented by quacks and mountebanks. To some extent she forced the society to become what they most despised: mountebanks and performers. Cavendish's reaction to Boyle's experiments was the same as if she were viewing jugglers and monstrosities at carnival or a country fair. Rather than being impressed with the achievements of the Royal Society, Cavendish reduced its most valued accomplishments to sideshow stunts. At the same time, her deportment and dress suggested that she herself was more worthy of admiration and wonder than anything the society could produce.

This was not the first time Cavendish used spectacle to establish her grandeur and authority, although past showcases had been fictional rather than real. In *Blazing World*, the fish-men subjects of the Empress, Cavendish's avatar, glide over the seas in battle to subdue the enemies of her native country—the land of E S F I. The Empress's appearance, resplendent in the star-stones and carbuncles of the Blazing World and the fire-stones of her worm-men, undermines and overcomes the enemy soldiers. After the battle, the princes of all the lands of E S F I gather to wait upon her, and she causes "a great admiration in all who were present, who believed her to be some Celestial Creature, or rather an uncreated Goddess, and they all had a desire to worship her." Showing herself "in her garments of Light," she is carried by the bird-men into the air, "and there she appear'd as glorious as the Sun." When she comes down to the sea again, she and her admirers are serenaded by "the most melodious and sweetest Consort of Voices, as ever was heard out of the Seas, which was made by the Fish-men; this Consort was answered by another, made by the Bird-men in the Air, so that it seem'd as if Sea and Air had spoke, and answered each other by way of Singing Dialogues, or after the manner of those Plays that are acted by singing Voices."[84]

Cavendish's visit to the Royal Society was a reenactment of the spectacle her other self in the Blazing World had already performed. Cavendish brought with her the Ferrabosca, a female member of a famous Italian singing family who had served the court for three generations. When, during her entrance into the experimental space of the Royal Society, Cavendish commanded the singer to "show her face and kill the gallants," we can hear the bird-men trilling in the background. The only music the Royal Society would hear, however, would be the laughter of the onlookers.

The members of the Royal Society never understood how they were the butt of Cavendish's joke. In February 1668, they asked the duchess to subscribe to their building fund. She declined the honor.[85] And in the dedication to her last book, *The Grounds of Natural Philosophy* (1668), "To All the Universities in Europe," she wrote, "All Books, without exception, being undoubtedly under your Jurisdiction, it is very strange that some Authors of good note, are not asham'd to repine at it; and the more forward they are in judging others, the less liberty they will allow to be judg'd themselves. . . . You are the Starrs of the First Magnitude, whose Influence governs the World of Learning."[86]

In fact, the Royal Society had hoped to preempt the authority of the universities by making Charles II its patron. It was a strategy that did not work. According to Pepys, in 1664 the king had spent "an hour or two laughing . . . at Gresham College. . . . Gresham College he mightily laughed at, for spending their time only in the weighing of ayre, and doing nothing else since they sat." By 1668, the king referred to the society as his "fous" or jesters.[87] Perhaps the king had lent his court singing lady to Cavendish to emphasize the foolishness of the new institution. Clearly, the society was having a public relations problem, and Cavendish's visit could only have increased the ridicule with which at least some members of the court and the public viewed the group.

In visiting the Royal Society, Cavendish was reified fancy turning the pursuits of experimental philosophy into a game. Her visit must be read like her books, a fusion of the rational and the fantastic. It perhaps represents Cavendish's conquest of her own fears about her work and her gender. Cavendish knew that in writing about natural philosophy she had crossed the boundary of what was acceptable to her sex. In order to defend herself, she appropriated and overturned the cultural categories of her time, arguing that uneducated women were particularly adept at fancy and at describing nature in a way the nonlearned could understand. She constructed her work and herself as a wonder: a writing woman become a female natural philosopher. Using herself as a foil, she ridiculed the program and pretensions of the Royal Society. Throughout her work, Cavendish used her unique role to integrate both popular and elite conceptions of the natural world and to challenge male hegemony in natural philosophy.

# chapter two

# Cavendish's Early Atomism

*I*t was no easy task to be a female natural philosopher or poet in seventeenth-century England. A preface to *Poems, and Fancies* expresses both Margaret Cavendish's self-doubt and her justification for writing: she wrote in verse, she explained, because "Errour might better passe there, then in Prose" and "Fiction is not given for Truth but Pastime." But she feared, "My Atomes will be as small Pastime, as themselves: for nothing can be less then an Atome. But my desire that they should please the Readers, is as big as the World they make; and my Feares are of the same bulk; yet my Hopes fall to a single Atome agen: and so shall I remaine an unsettled Atome or a confus'd heape, till I heare my Censure. If I be prais'd, it fixes them; but if I am condemn'd, I shall be annihilated to nothing: but my Ambition is such, as I would either be a World, or nothing."[1]

The world Cavendish intended to create was her book. In the last of her prefatory material, she pleads with her readers that if they do not like her book, "Disturb her not, let her in quiet dye."[2] The book itself is figured as female and indeed is Cavendish's progeny: "Condemne me not for making such a coyle/About my Book, alas it is my Childe."[3] Cavendish often thought in correspondences: thus, she is an atom, a creator, and a mother; likewise, her book is a world, a creation, and a child. The legitimacy of her work results from the author being, in some sense, both the form and matter of her work, both the progenitor and the progeny. In early modern England, female imagination was supposed to affect the appearance of the child; Cavendish wanted her conceptions to be beautiful and not monstrous.[4]

Her children took many forms, showing her readers many worlds. In *Poems, and Fancies*, creation is conceived as the product of a female Nature. In its companion piece, *Philosophicall Fancies*, the malleability of matter can produce other worlds where people are made of flowers or iron and where rocks and animals may possess reason. All of Cavendish's worlds are inhabited, so in *Poems, and Fancies* atoms are analogized to fairies. In her later works, there will be realms peopled by hybrid beings, worlds existing both below and above the earth, in the oceans and in space. Inspired by the new lands and the new science discovered in the sixteenth and seventeenth centuries, and by poets' and philosophers' perceptions of the new realities, Cavendish's fancy produced multiple books and multiple possibilities. In creating other worlds, Cavendish defined a space where her fancy could discover and envision nature.

It is not surprising, therefore, that Cavendish's first work began with a cosmogony based on a material creator using living matter to construct the world. Most commentators, including myself in an earlier discussion, have followed Robert Kargon in his interpretation that Cavendish "expounded an Epicurean atomism at once so extreme and fanciful that she shocked the enemies of atomism, and embarrassed its friends."[5] Epicurean atoms are simply material; they possess motion and figure but not life. Both moderns and ancients viewed such a conception with repugnance. Pierre Gassendi, Newcastle's guest in 1647 and Thomas Hobbes's friend, spent many hundreds of pages in his *Syntagma Philosophicum* (1655) trying to rehabilitate this ancient philosophy and integrate it with Christianity.[6] Hobbes was often accused of being an Epicurean and, once again, was doubtful intellectual company for Cavendish to keep.

But Cavendish was not a classic Epicurean. By 1653 she had already embraced a vitalist theory of matter, which becomes explicit in *Philosophicall Fancies*, which she intended to publish in the same volume as *Poems, and Fancies*. Cavendish's vitalism was as idiosyncratic as the rest of her philosophy; she reconfigured chemical ideas of vital heat and seminal principles into a material philosophy that credits matter not only with life and self-movement but also with self-consciousness and thought. And, in addition to her embrace of vitalism, Cavendish deviated from the Epicurean norm by integrating theological motifs into *Poems, and Fancies* and all her scientific works. Cavendish's Epicurean tendencies were reflected in her ties to Gassendi and his English disciple Walter Charleton, both trying to make Epicureanism acceptable to a Christian conscience. Charleton was perhaps the most immediate influence on Cavendish, but her articulation of a living universe was all her own.

Cavendish included a cosmogonic poem at the beginning of *Poems, and Fancies*, and she included theological verses near the end of the atomistic poems. She advocated a kind of negative theology: we can know only that God is not like us. Indeed, all of her works reflect a fideism so pervasive that it approaches eighteenth-century deism. It is the theological counterpart of her skepticism. Fideism, a common religious stance among skeptics, was the belief that God is unknowable and therefore cannot be understood by human reason. Fideism functioned as a corrective to the enthusiasm in religion that had produced religious war in Europe and England for more than a century before Cavendish wrote. Fideists and skeptics argued that since we can believe in God only through faith, and that there is no way for a human being to judge what understanding of God is true, we may as well adhere to the tradition into which we were born.[7] This kind of a religious attitude cohered well with Cavendish's essentially conservative attitude toward religion and political authority, while at the same time allowing her philosophy equal validity with other probabilistic accounts of nature. In *Poems, and Fancies*, she states her unwillingness to discuss the nature of God: "For it were too great a Presumption to venture to Discourse that in my Fancy, which is not describeable. For God, and his Heavenly Mansions, are to be admired, wondred, and astonished at, and not disputed on."[8]

Fideism remains constant throughout her works. In her *Orations* (1662), Cavendish wrote, "Whatsoever is Infinite and Eternal, is God, which is something that cannot be Described or Conceived; not prescribed and bound, for he hath neither beginning nor ending."[9] In several of Cavendish's works, God is described as "the diatical center" from which all things flow and return but is unknowable.[10] Such an attitude was condemned by the Catholic Church, and the Protestant faiths viewed it with suspicion.[11] It was particularly problematic for a philosopher who grounded her material philosophy on atomism. The gods Epicurus had envisioned in antiquity were completely absent from the governance of an eternal universe—it would cause them too much trouble. The God Cavendish pictured in her later works delegates authority to a female deity, Nature, who is better fitted to generate a living universe. This unknowable God was an absent God in much of *Poems, and Fancies*, and this disappearance can lead to the conclusion that Cavendish's theology is nature worship at best and atheism at worse.

*Poems, and Fancies* can be divided into two sections. The first consists of her atomistic poems, and the second, meditations or "fancies" about morality, government, psychology, and anything else that struck her fancy—including fairies. Nevertheless, there is a consistency to the book's argument. What may seem like a jumble of scientific and fantastic speculation obeys an inner logic. Cavendish's description of fairyland is actually a recasting of her initial atomistic theory. The atomism itself is presented in the context of a materialist metaphysics in which a corporeal, gendered creator, called Nature, directs material principles, also gendered, to establish a world and all that is in it using atoms. While her world recalls that of Epicurus—the atoms are infinitesimally small bits of the same matter, distinguished only by shape and size, whirling around in a void—the atoms' character is profoundly different. In Cavendish's universe, all beings, both creator and created, are in some sense alive. Motion, reified sometimes as male and sometimes as female, ultimately is the most active force in the universe, endowing matter and its concretions with life. By the end of Cavendish's poems, motion is deified; by the end of her fancies, moving atoms are identified with fairies. Matter and motion, the two essential properties of the mechanistic universe Cavendish encountered as an exile in France, are transformed into the principles of a universe humming with life, at every level of existence.

### POEMS, AND FANCIES: BEGINNINGS AND ENDINGS

*Poems, and Fancies* was dedicated to Cavendish's brother-in-law Sir Charles Cavendish, with whom Cavendish had much "delightfull conversation" while the two journeyed to England in 1651 in an attempt to gain some of the moneys from Newcastle's confiscated estates. However, when Margaret Cavendish went before a parliamentary committee to make her plea, she became mute: "I whisperingly spoke to my brother to conduct me out of that ungentlemanly place, so without speaking to them one word good or bad, I returned to my Lodgings, &, as the Committee was the first, so was it the last, I ever was at as a petitioner."[12] Part of the motivation for Cavendish's unusual public appearances in the later part of her career, including her visit to the Royal Society, may have been her effort to negate this earliest failure in the role of a public person. She often characterized herself as "bashful." But we will see that in both her life and her works Cavendish would command, not implore.

Although Parliament might have silenced Cavendish, Sir Charles supported her. The dedication of *Poems, and Fancies* captures the poignancy of their friendship: "But certainly your Bounty hath been the Distaffe, from whence Fate hath Spun the thread of this part of my Life, which Life I wish may be drawne forth in your Service."[13] This poetical conceit prefigures her own hermaphroditical role as a mother/creator whose actions combine female generation with male authority.

Contemporaries often commented on the sweetness of Sir Charles's disposition, and it seems he did not fail to encourage his sister-in-law, in both her natural philosophy and her fancy.[14] At the very end of *Poems, and Fancies*, Cavendish wrote,

> Sir Charles into my chamber coming in,
> When I was writing of my Fairy Queen; I pray, said he, when Queen Mab you doe see,

Present my service to her Majesty: . . . In whispers soft I did present
His humble service, which in mirth was sent.[15]

Sir Charles's mild teasing clearly is recounted to underscore the frivolity of Cavendish's labors; she presented *Poems, and Fancies* as a harmless pastime, something to occupy her idle hours. In an apologia dedicated "To Naturall Philosophers," Cavendish fashioned her work as a divertissement: "I cannot say, I have not heard of Atomes, and Figures, and Motions and Matter; but not throughly reason'd on: but if I do erre, it is no great matter; for my Discourse of them is not to be accounted Authentick: so if there be any thing worthy of noting, it is a good Chance; if not, there is no harm done, nor time lost. For I had nothing to do when I wrot it, and I suppose those have nothing, or little else, to do that read it."[16]

Cavendish was both dismissive and arrogant about her work. She knew that most readers would not be interested in her version of atomism: "I desire all that are not quick in apprehending, or will not trouble themselves with such small things as Atomes, to skip this part of my Book, and view the Other, for fear these may seem tedious: yet the Subject is light, and the Chapters short." For readers, perhaps including her husband, who expected her book to be amusing, she wrote, "I wish heartily my Braine had been Richer, to make you a fine Entertainment." Nevertheless, she concluded, serious philosophers would recognize the worth of her atomistic musings:

let my Atomes to the Learned go
If you judge, and understand not, you may take
For Non-sense that which learning Sense will make.[17]

Whether either a learned or popular audience would take Cavendish's offering as a kind of joke with important hidden meaning is open to question. That she intended her scientific poetry as a kind of ploy or tactic to allow her to infiltrate male preserves of knowledge is also possible. But the overall impression of her first two works dealing primarily with scientific topics—*Poems, and Fancies* and *Philosophicall Fancies*—is that they were spontaneous and that she viewed their lack of order and discipline as an inevitable consequence of unfettered fancy, which to her mind was not such a bad thing. She shared, after all, this proclivity with Nature herself:

For Nature's unconfin'd, and gives about
Her severall Fancies, without leave, no doubt.
Shee's infinite, and can no limits take,
But by her Art, as good a Braine may make.[18]

Cavendish's natural philosophy in her early works is obscure and inconsistent and clearly confirms her claim that she had not read or studied other philosophies but merely absorbed her knowledge from her male relatives. She excuses herself both because of her sex, which values fancy that "goeth not so much by Rule, & Method, as by Choice," and because of her lack of knowledge of philosophic terms: "Neither do I understand my owne Native Language very well; for there are many words, I know not what they signifie."[19] Clarity is sacrificed to metaphor and further undermined by what Cavendish acknowledged later was a very inaccurate printing.[20]

*Poems, and Fancies* did not receive the reception Cavendish wanted. Some readers did not believe it was her own composition, or they disparaged her work as foolish and absurd.[21] Her contemporary Dorothy Osbourne dismissed Cavendish's poems as "ridiculous," and some 250 years later Virginia Woolf deplored the effect natural philosophy had on Cavendish's poetry: "Under the pressure of such vast structures, her natural gift, the fresh and delicate fancy, which had led her in her first volume to write so charmingly of Queen Mab and fairyland, were crushed out of existence."[22]

What readers then and now have failed to grasp is that Cavendish's decision to mix reason and fancy was integral to her system: her natural philosophy could be understood only by combining the philosophic and the fantastical. While she might present her work in the guise of amusement, even in her first published works serious philosophic and theological concerns affected her ideas, which she expressed in poetry and fables. Such a fusion of philosophy and fancy in her works reflected the operation of the mind, or what she will call later rational matter. In *The Worlds Olio*, published in 1655 but written in 1653, Cavendish describes the actions of the brain once it has been presented with sense impressions, which "then the brain cuts and divides them," making "thousands of several figures, and these figures are those things which are called, imagination, conception, opinion, understanding, and knowledge, which are the Children of the brain."[23]

*Poems, and Fancies* begins with a cosmogonic myth describing a female-gendered Nature calling a council of her advisers, the natural principles, on how to create a world: "Motion was first, who had a subtle Wit,/And then came Life, and Forme, and Matter fit."[24] Nature is clearly in charge; the most fundamental aspect of her being is to command, while the other principles must obey: "Besides it is my nature things to make,/To give out worke, and you directions take." She is the first of the many women who give orders (and make order) in Cavendish's fiction. Her motivation is the desire for adoration:

First Nature spake, my Friends if we agree,
We can, and may do a fine Worke, said she,
Make some things adore us, worship give,
Which now we only to our selves to live.[25]

Nature needs creation in order to be worshiped. A self-referential self-regard is not sufficient for a creative force: she must make nature—the material realm—in order to supply the audience her self-esteem demands. Nature craves attention, and she thinks her ministering principles should also be worshiped; in later works many of Cavendish's projected selves will be worshiped. Nature is neither self-sufficient nor autocratic—she is not an omnipotent deity. She is essentially unlike the Judeo/Christian God, who is markedly absent in this story of creation. Whether God made Nature, and his relationship to her, is a theme that Cavendish will consider only in her later works.

One of the difficulties in understanding Cavendish's natural philosophy is the ambiguous identity of nature, which is variously seen as a personified force (referred to with a capital N in this book) or the sum total of matter that constitutes the world (referred to with a lowercase n). Sometimes Nature is viewed as distinct from her material aspect, and sometimes she is subsumed within it. The dynamic relationship between Nature the deity and nature the material is indicated in Cavendish's first work, *The Worlds Olio*. Reflecting on how much she had learned from Sir Charles, Cavendish concludes, "Though I do not write the same way you write, yet it is like Nature which works upon Eternal matter, mixing, cutting, and carving it out into several Forms and Figures."[26] In other words, Cavendish knew that she was reshaping the ideas she had received from her brother-in-law; it is impossible not to think of Nature and Cavendish as cooks or housewives working on dough and turning it into bread. Cooking as a methodology becomes a heuristic for the natural principle, the writer, and the cook, all creators in their own spheres of being.

And just as the cook could not be a cook without the ingredients for her bread, likewise Nature needs the matter she uses: "For had not Nature Matter to work upon, She would become Useless; so that Eternal Matter makes Nature work, but Nature makes not Eternal Matter." In this account, the subservient role of Nature in relation to her creation ultimately reverses the status of the two principles of being, with Nature becoming "a labouring servant," and "Eternal Matter" possessing "Spirit and Motion, which is Life and Knowledge."[27] A clearly female principle becomes subservient to the life she forms. In this way, Cavendish implied her own dependence on Sir Charles and the ideas he had taught her but her ultimate independence as both the creator and creation of her universe.

In *Poems, and Fancies* Cavendish imagines a multitude of personified forces affecting the creation. The fates and destiny exist—whether by the command of Nature or independent of her is left unclear—but they are used by Nature to undermine the power of inconstancy and fortune. Just as she thinks in correspondences, Cavendish also often thinks in contrarieties: destiny's chains confine fortune's wheel, while wet and dry and hot and cold—also in some preexistent form—work to form a kind of cosmic balance. It is in this time and space that Motion, Nature's prime agent, begins to work.

Motion's first task, Nature instructs, is to create light in order to prevent the domination of vacuum and darkness, two other active principles possessing the power to cover Nature's world. Obviously, this is a universe seething with various powers and their conflicts. Motion is unable to carry out Nature's command except by employing her other lieutenants: Matter, Figure, and Life:

Alas, said Motion, all paines I can take,
Will do no good, Matter a braine must make;
Figure must draw a Circle, round and small,
Where in the midst must stand a Glassy Ball,
Without Convexe, the inside a Concave,
And in the midst a round small hole must have,
That Species may passe, and repasse through,
Life the Prospective every thing to view.[28]

To say that this passage is opaque is an understatement. What does Cavendish mean when she writes that Matter must make a brain? Perhaps she imagines the whole or some of matter being like a brain, perceptive and sensate—an idea more clearly expressed in later works. Motion itself cannot function unless conjoined with matter, and matter itself must have a shape. In this case, it appears that the world (or brain) that motion and matter make is solid and round and that species—material beings of some kind—will eventually enter the world through an opening in the world. It's possible that Cavendish meant "species" to indicate individuated members of genera in the Aristotelian sense. Alternatively, she may have associated "species" with the Epicurean idea of films of objects that impinge on the eye, creating perception. One of the seventeenth-century meanings of "species," according to the *OED*, was "the outward appearance or aspect, the visible form or aspect, of something, as constituting the immediate object of vision."[29] It is likely that Cavendish was equating the "species" with the atoms that she will shortly describe at length. These species are observed by Life, whose role in all of this is ambiguous. She observes the process of world making, but whether she will become an intrinsic part of the matter that composes the world is left unclear.

However, the vitality of matter in motion is affirmed immediately in Cavendish's cosmogonic poem when Cavendish identifies Death as the primary enemy of creation: "Alas, said Life, what ever we do make, / Death, my great Enemy, will from us take." The three other principles equally lament the dissolution that Death inevitably brings, and they eventually conclude that it might be best just to let Death have his way:

> Then Motion spake, none hath such cause as I
> For to complaine, for Death makes Motion dye.
> 'Tis best to let alone this worke, I thinke.
> Saies Matter, Death corrupts, and makes me stinke.[30]

Death is the enemy of Life and, according to Cavendish, is the ultimate nemesis of Nature herself, who fears, "If we let Death alone, we soone shall finde, / His wars will make and raise a mighty power / If we divert him not, may us devoure."[31] There is a kind of existential angst in the reaction of Motion and Matter to death. They seem to be unable to prevent this opposing force from corrupting their work. But Nature refuses to allow them to wallow in despair. She orders the construction of the universe:

> First Matter she brought the Materialls in,
> And Motion cut, and carv'd out every thing.
> And Figure she did draw the Formes and Plots,
> And Life divided all out into Lots.
> And Nature she survey'd, directed all,
> With the foure Elements built the Worlds Ball.[32]

Creation culminates in Nature's decision to make man, "not like to other kinde, / Though not in body, like a God in minde." Nature endows man with "With Knowledge, Understanding, and with Wit," while the four principles are given the job of fashioning man's body and endowing it with passion or sensation, an attribute man shares with other created beings. What man does not share is his shape: he is unique in being upright. He also is profoundly unique in another way: man possesses freedom of will and freedom from death, at least spiritually, and perhaps materially. Nature states, "But yet the Minde shall live, and never dye; / We'le raise the Body too for company.[33]

A creative force that has the power to resurrect the soul and body is perhaps too close to Christian heresy for Cavendish to pursue. She abruptly concludes her fancy about Nature and her lieutenants and begins her atomistic poems. Nature's place is taken by self-moving atoms that direct their own activities:

> Small Atomes of themselves a World may make,
> As being subtle, and of every shape:
> And as they dance about, fit places finde,
> Such Formes as best agree, make every kinde
> And thus, by chance, may a New World create:
> Or else predestinated to worke my Fate.[34]

In this poem, the autonomy of atoms obviates the necessity of a personified nature. The atoms incorporate the principles of figure and motion, and they seem to possess consciousness, the ultimate characteristic of life. Their dance may simply be a random tumbling that produces worlds by chance, or it may determine creation, including human life.[35] Cavendish thus allows the action of the atoms to embrace freedom and determinism, the most contested categories of theological and scientific thought during her time.

The dancing atoms constitute the material substratum of the universe. Although the matter in the different atoms is the same, they are differentiated by the forms their internal motion creates. The different shapes of atoms create another form of materiality, the elements. Thus, square atoms constitute earth, round atoms make up water, long atoms compose air, and sharp atoms constitute fire. The pointed atoms are associated with life, "All pointed atoms to Life do tend, / Whether pointed all, or at one end," but whether this means that the other atoms are inert is difficult to determine. It may be that life is a by-product of atomic concretions, as another poem suggests: "The Cause why things do live and dye, / Is as the mixed Atomes lye."[36] Or perhaps, Cavendish suggests, motion is the cause of life: "So Life doth only in a Motion lye. / Thus Life is out, when Motion leaves to bee."[37] In some cases, life expectancy is related to how the differently figured atoms cohere. Thus, vegetables decay much more quickly than minerals, while "In Animals, much closer they are laid, / Which is the cause, Life is the longer staid."[38]

The various permutations and movements of these atoms, and the void, produce all the variety of form and change we find in the world: mineral, vegetable, animal, astronomical, and meteorological. And so, Cavendish concludes,

> Thus Vegetables, Minerals do grow,
> According as the severall Atomes go.
> In Animals, all Figures do agree;
> But in Mankinde, the best of Atomes bee.[39]

These better atoms work together to make the human constitution and psychology. Their motion in the brain constitutes our understanding and emotions, as well. Their harmony produces health; their disharmony, disease.

Cavendish wanted to make sure her reader understood all the properties of atoms, and so in the middle of her discussion of the elements, she repeats their characteristics, in a poem called "A World made by foure Atomes." She writes, "This I do repeat, that the ground of my Opinions may be understood. . . . But the foure Atomes meet, and joyne as one. And thus foure Atomes the Substance is of all; With their foure Figures make a worldly Ball."[40]

Somewhere in her discussion, Cavendish had ceased to think of her speculations as fancies. She wanted her ideas to be taken seriously. Increasingly, the more Cavendish wrote, the more she wanted other thinkers to recognize the possible truth of her materialist understanding of nature. But in this text, Cavendish explicitly retreated from a cosmos that is determined by its material constituents. Cavendish included a poem, "The Motion of Thoughts," in *Poems, and Fancies* in which she described a beatific vision of God. While she was walking, her thoughts were wandering: "At last they chanc'd up to a Hill to climb, / And being there, saw things that were Divine." Her vision is of a glorious light, in constant motion, "This Light had no dimension, nor Extent, / But fil'd all places full, without Circumvent; / Alwaies in Motion, yet fixt did prove."[41] This prime motion is unchanging and self-contained:

> For the first Motion every thing can make,
> But cannot add unto it selfe, nor take.
> Indeed no other Matter could it frame,
> It selfe was all, and in it selfe the same.
> Perceiving now this fixed point of Light,
> To be a Union, Knowledge, Power, and Might;
> Wisdom, Justice, Truth, Providence, all one,
> No Attribute is with it selfe alone.
> Tis its own Center, and Circumference round,
> Yet neither has a Limit, or a Bound,
> A fixt Eternity, and so will last,
> All present is, nothing to come, or past,
> A fixt Perfection nothing can add more,
> All things is it, and It selfe doth adore.[42]

The Divine Light is a circle, like the world that Nature and her ministers created in Cavendish's first poem, or the divine diatical center described in later works. Like that world, the center and circumference of the Divine Light meet in a kind of mystical union, which is unbounded and perfect. Whether it is material is not addressed. In the story of creation as we have seen, Nature feared, "Vacuum, and darknesse they will domineere, / If Motions power make not Light appeare . . . My only Childe from all Eternitie."[43] In the cosmogony, motion makes light; in the theological poem, light is prime motion. Like Aristotle's first mover, it needs only itself and is content to adore itself. Such self-sufficiency seems distant from Christianity, but the attributes of this divinity make it more amenable to a Christian reading. It shares with the Christian God unity, omnipotence, omniscience, and virtue, as well as providence. Nature had care for her creation, and particularly for man, but needed the world in order to be adored. The Divine Light has no such need, and consequently its wisdom, justice, truth, and providence must be, at least in some sense, altruistic. Cavendish finally perceives that her thoughts share in the divinity of God. In the earlier poem, Nature had made the mind eternal like herself, but in this poem, Cavendish's thoughts are part of the Divine Light itself:

> My Thoughts then wondring at what they did see,
> Found at the last themselves the same to bee;
> Yet was so small a Branch, perceive could not,
> From whence they Sprung, or which waies were begot.[44]

Divinity overflows from the Divine Light. Cavendish added in a marginal note, "All things come from God Almighty." "All things" apparently included her—not just as the product of God but as somehow organically related to him. At this point, however, she pulled back from the pantheistic implications of her poem. Instead of the contemplation of the divine leading to knowledge, it leads only to joy, but joy cannot be understood in its heavenly ecstatic form: "Some say, all that we know of Heaven above, / Is that we joye, and that we love, / Who can tell that?"[45]

In fact, she concludes, the resurrection of the body is so transformative that we may, as Saint Paul suggests, "see God with our Eyes," but the change in body and mind may be so profound, "As if that we were never of Mankind, / And that these Eyes we see with now, / were blind." Thus, her original beatific vision is compromised by the promise of divine sight in heaven, and so she concludes, fideistically, "yet cannot all the Wise, and Learned tell, / What's done in Heaven, or how we there shall dwell."[46]

Cavendish's mystical vision ends in skepticism rather than gnosis. There is no element of a personal relationship to God or Christ in any of her works. Nevertheless—as with any seventeenth-century natural philosophy—religious themes permeate her discussion.

## CAVENDISH AND THE ATOMISTS

If Cavendish was an Epicurean, the religious elements in *Poems, and Fancies* are more than strange. Epicurus formulated his atomic theory in order to remove the fear of death and the fear of the arbitrary actions of the gods. Since matter simply reverts to its original atoms when a human being dies, there is no need to be terrified by death. Likewise, whatever gods there may be are completely uninvolved in the world—they exist only to pursue their own tranquility without the bother of providentially ordering and maintaining the universe. Not surprisingly, Christian authors rejected Epicurean moral and natural philosophy until it was rehabilitated by the French priest and mechanical philosopher Pierre Gassendi. In the early seventeenth century, Gassendi argued that matter was impressed with motion in the beginning and maintained by God's direct action always. He also had no doubt about the Christian belief in the immortality of the soul.[47] Cavendish had met Gassendi at her husband's salon in the late 1640s, when his Epicurean project was well advanced, and undoubtedly heard discussions of his ideas. Her inability to read Latin meant that her knowledge of his work was mediated through others.

If one looks more closely at her theological beliefs, her distance from Gassendi becomes very apparent. However, she may have become intrigued with the idea of an "atom," a subsensory particle of matter, because of the interest her husband and Sir Charles took in Gassendi's neo-Epicureanism. Throughout the later 1640s, Sir Charles often met and spoke with Gassendi, about whom he wrote, "I am much beholding to him for his visits and freedom of discourse with me." In 1650, Sir Charles was perusing a copy of Gassendi's *Animadversiones in decimum librum Diogenis Laertii* (1649), which contained a summary and critique of ancient atomism: "I have rather have turned over the leaves and onlie read the sum of Epicurus His Philosophy than anie otherwise, but as far as I can guess my worthie friend Gassendes hath both maintained and opposed Epicurus where he ought most excellentlie."[48]

Just at the time Sir Charles was reading Gassendi, he and Cavendish traveled to England. Once there, Cavendish became despondent because of her separation from her husband, leading her to consult Walter Charleton (1619–1707), a royalist physician who shared the contemporary interest in things scientific and later was a member of the Royal Society. After Cavendish published her early scientific books, Charleton reported to the Cavendishes the common disbelief that Cavendish was the author of her own works.[49]

At the time Charleton met Cavendish, he advocated the Paracelsian natural philosophy of Van Helmont, but he soon came to support Descartes and especially Gassendi, probably as a result of meeting the royalist exiles.[50] His book *The Darkness of Atheism Dispelled by the Light of Nature* (1652) contains a selective and not very accurate translation of Gassendi's *Animadversiones* and includes a description of the French philosopher's entire system. The *Physiologia Epicuro-Gassendo-Charletoniana* (1654) is for the most part a translation of the physical section of the *Animadversiones*. Charleton wanted to defend Gassendi's philosophy from any imputation of atheism by showing how atomism, if properly understood in its Gassendist form, affirms God's existence and action. In particular, Charleton argued that the order and design of the universe testify to God's creative action and the Epicurean belief in the eternal, fortuitous action of the atoms must be incorrect. With these revisions, Charleton concluded, atomism is a likely explanation for the nature of the universe: "I have never yet found any justifiable ground, why Atoms may not be reputed Mundi materies, the Material Principle of the Universe, provided that we allow, that God created the first Matter out of Nothing; that his Wisdome modelled and cast them into that excellent composure or figure, which the visible World now holds; and that ever since, by reason of the impulsion of their native Tendency, or primitive impression, they strictly conform to the laws of his beneplacuits, and punctually execute those several functions, which his almighty Will then charged on their determinate and specifical Concretions."[51]

The principles are the same as Cavendish's—matter, figure, and motion constitute the universe—but the conclusion is very different. To Charleton, no principle exists independently of God. God uses no ministers—either Nature or her lieutenants—to create the universe. All things prove the absolute power of God and are contingent on his will.[52] Motion is used by God to carry out his law in the universe. The question of the nature of the atoms, whether they are lifeless or animate, remains open.[53]

Charleton's main objection to Epicurean atomism was that in its theory of the creation of the world, it substituted the chance meetings of atoms for the eternal providence of God. Such a view is a romance or a fable, which "strikes the nosethrills as wel of the meer Natural man, as the Pious, with such infectious stench, that nothing but the opportunity of confutation can excuse my coming so neer it."[54] If Charleton was so appalled by the fables of Epicurus, how could he have reacted to Cavendish's atomistic fable, which begins with a personified nature and ends with a god of motion? In fact, Charleton had no choice but to applaud Cavendish's efforts—she was, after all, the wife of an earl, even though an exiled one— but his praises reflect his uneasiness. In a 1654 letter, he wrote to her, "Whenever my own Reason is at a loss, how to investigate the Causes of some Natural Secret or other, I shall relieve the Company with some one pleasant and unheard of Conjecture of yours. So that by reading of your Philosophy, I have acquired thus much of advantage; that where I cannot Satisfy, I shall be sure to delight: which is somewhat more than I dare promise from any other Discourses . . . in so much as they generally leave the Mind in a kind of Anxiety and Regret, when ever they fail to afford it Satisfaction."[55]

By 1651, Thomas Hobbes was also in England, having just published his *Leviathan*. Although Hobbes had been a client of her husband's, Cavendish claimed that the great materialist had refused her offer to dine.[56] Possibly, Charleton was referring to Hobbes when he expressed anxiety over the writing of other natural philosophers and had warned Cavendish of the danger of being associated with the heterodox thinker. Did Cavendish insert a cosmogony and ecstatic vision as the frame for her atomism because she wanted to ensure that her atomic poems would be viewed as simply the fabulous conjectures of a female mind? Did she use them as a means of avoiding the imputation of heresy often associated with atomism and materialism, against which her fellow atomist Charleton fought so furiously? Did Cavendish use Nature and her agents as a subterfuge to deflect attention from the heterodoxy of her work? Or did she simply want to delight her readers with her entertaining version of natural philosophy, an aim not inconsistent with her husband's view of the scientific enterprise?

Perhaps the answer is all of the above, although Cavendish was probably not worried about the apparent contradictions between her ideas and Christian theology. Her audience was not theologians but ordinary readers, who—whatever their social class—accepted allegory and mythologizing in literature.[57] As Charleton himself recognized, "the Fictions of Poets and Romancists" were imaginary, and their purpose was not the discovery of truth but the delight of the reader. Consequently, he concluded, "Man can run but little hazard of his Judgement, who shall affirm, that your Suppostion of Fayeries in the Brain, and of our Thoughts being their Consults and Suggestions; and your Opinion that Fayeries digging for Stones in the Quarries of the Teeth, to repair their decay'd Tenements in the Head, is the Cause of the Tooth-ach; are as worthy the hearing, as the most solid demonstrative Theory of any Philosopher whatever; insomuch as they yield both as high and lasting a Delight as that."[58]

For Charleton, atoms and fairies, as Cavendish described them, share the same imaginative space: they are fables, but not like the fables of Epicurus he deplored. Charleton was correct in perceiving the close association of Cavendish's natural philosophy with her tales of fairies but incorrect in thinking that because the descriptions of fairies were fancies Cavendish viewed them as untrue.

## ATOMS AND FAIRIES

Just as there is unity between Cavendish's atomistic poems and their supporting frame, likewise there is a confluence between her poems and her fancies, particularly the works of imagination that deal with fairies. At the conclusion of the atomistic poems, Cavendish seems to segue into fantasy. Beginning with her skeptical premise that the senses cannot penetrate into the interior of nature, Cavendish argued, "So in this World another World may bee,/That we do neither touch, smell, heare, see." We cannot see the atoms in the air, but we know light exists. Likewise,

And whatsoever can a Body claime,
Though nere so small, life may be in the same,
And what has Life, may Understanding have,
Yet be to us buried in the Grave.
The[n] probably may Men, and Women small,
Live in the World which wee know not at all.[59]

The vision of layer after layer of animate, inhabited worlds within atoms continues in the next poem, where Cavendish informs the reader that Nature, here personified as a creative source, makes many worlds because of curiosity. These creations teem with life and are composed of the four different kinds of atoms explained earlier in her text. In fact, the difference between atoms and infinitesimal beings collapses in the course of the poem:

For Creatures, small as Atomes, may be there,
If every Atome a Creatures Figure beare,
If foure Atomes a World can make, [marginal
note: "As I have before shewed they do, in
my Atomes"], then see
What severall Worlds might in an Eare-ring be. . . .
And if thus small, then Ladies well may weare
A World of Worlds, as Pendents in each Eare.[60]

Cavendish then describes this world in great detail, including its flora and fauna and its society and government. The tiny folk living in a lady's earring anticipate another diminutive race, one that will be described much later in her book, not as part of the "poems" but as a "fancy." We are asked,

> Who knowes, but in the Braine may dwel
> Little small Fairies; who can tell?
> And by their severall actions they may make
> Those formes and figures, we for fancy take.[61]

In Cavendish's worldview, atomism has been mythologized. In fact, the fairy kingdom itself, with all the pertinent detail about flora and fauna and society and government, is ruled by the fairy queen: "There Mab is queen of all, by Natures will,/And by her favour she doth govern still."[62] Nature has delegated the rule of fairyland to Queen Mab, and within her kingdom her tiny subjects "do keep just time and measure,/All hand in hand, a round, a round,/They dance upon this Fairy ground."[63] Fairies are the functional equivalent of the dancing atoms described in the first section of Cavendish's text; essentially atoms and fairies are they same—their motion causes thought and feeling. Cavendish entitled one poem "The Fairies in the Braine, may be the causes of many thoughts," which can only recall her earlier description of mental action: "So Wit, and Understanding in the Braine, Are as the severall Atomes reigne."[64] More important, fairies seem to be the same as the essence of life. In the atomistic poems, we are told, "For Animall Spirits, which we Life do call,/Are onely of the sharpest Atomes small," while the poems on fairies concludes, "Those Spirits which we Animal doe call,/May Men, and Women be, and Creatures small."[65]

Was Cavendish jesting? Perhaps. The fairies don't live in the same part of the book as Nature and the atoms. But Cavendish was not alone in her association of fairies and atoms. She is anticipated by Shakespeare, who in *Romeo and Juliet* gives this image to Mercutio:

> O then I see Queen Mab hath been with you.
> She is the fairies' midwife, and she comes
> In shape no bigger than an agate-stone
> On the forefinger of an alderman,
> Drawn with a team of little atomi
> Over men's noses as they lie asleep.[66]

Tiny beings pull Mab's chariot. They are an example of what Katharine Briggs calls "The Fashion for the Miniature" in Elizabethan and Jacobean literature, especially among the poets who followed Shakespeare in the early seventeenth century.[67] Cavendish was an admirer of Shakespeare and very likely was aware of

this passage.[68] Reginald Scot makes a similar reference in his witch-hunting man-
ual, the *Discoverie of Witchcraft* (1584), which was probably unknown to Cavendish. Scot describes a seeker after second sight, who sees "a Multitude of Wights,
like furious, hardie Men, flocking to him hastily from all Quarters, as thick as
Atoms in the Air."[69]

The image of dust motes as dancing atoms goes back to Lucretius, the Roman
poet who originally described Epicurean atomism.[70] He wrote,

> Observe, for instance, when the sunlight with its rays
> Penetrates and pours into a darkened room.
> Many minute motes you will see, and they will mingle
> In many ways through the air, dancing in the sunbeam, struggling and fighting
>     as if in an eternal battle,
> Bounding and rebounding together frequently;
> And from this you may imagine how the atoms
> Are always buffeted through unending space.[71]

Cavendish had no access to the text of Lucretius, although Emma Rees has
argued recently that her choice of genre, the philosophic poem, was influenced by
his model.[72] Rather, her access to this metaphor came through more recent literary
references: John Donne used the phrase "like those atoms swarming in the sun,"
and the writer of masques, John Sadler, described motes personified as ladies dart-
ing from the rays emitted by a queen, played in his masque by Henrietta Maria,
whom Cavendish would later serve as a maid-in-waiting.[73] Even if she had known
Lucretius, his description of inert particles making confused creations would have
been inimical to her. Her atoms are like fairies, not dust.

The unitary characteristic of atoms, fairies, and dust motes is smallness, as
well as a proclivity for dancing. The notion of fairies as small, material beings is
common to both popular and elite culture in the sixteenth and early seventeenth
centuries, and as Briggs points out, it is almost impossible to distinguish these elements in the literature of that period.[74] In a 1668 play, Cavendish refers to "to the
Queen of the Fairies, which is an old Wives-tale."[75] Cavendish incorporates all
the folkloric associations of fairies into one of her poems about Queen Mab:

> Her pastime onely is when she's on earth,
> To pinch the Sluts, which makes Hobgoblin mirth:
> Or changes children while the nurses sleep,
> Making the father rich, whose child they keep.
> This hobgoblin is the Queen of Fairies Fool,

Turning himselffe to Horse, Cow, Tree, or Stool;
Or any thing to crosse by harmlesse play,
As leading Travellers out of the way,
Or kick downe Payls of Milk, causing Cheese not turn,
Or hinder Butter's coming in the Churne.[76]

The mischievous nature of fairies is a staple of folklore, particularly as fairies lost any association with witchcraft in the seventeenth century. Cavendish still blamed them for unusual domestic problems but viewed their interventions as harmless; even changelings only bring riches to their fathers. The recreation of the fairies is dancing:

Where this Queen Mab, and all her Fairy fry,
Are dancing on a pleasant mole-hill high:
With fine small straw-pipes sweet Musicks pleasure,
By which they do keep just time and measure.
All hand in hand, a round, a round,
They dance upon this Fairy ground.[77]

The fairies are tiny dancers: they are small material beings who keep time and measure. So are the atoms:

Atomes will dance, and measure keep just time;
And one by one will hold round circle line,
Run in and out, as we do dance the Hay.
Crossing about, yet keepe just time and way.
While Motion, as Musicke directs the Time:
Thus by consent, they altogether joyne.
This Harmony is Health, makes Life live long.
But when they're out, tis death, so dancing's done.[78]

There is nothing more purposeful than the dance, which is a metaphor for order, whether in the celestial spheres, the terrestrial orb, or the material world. As long as beings dance, they possess life and health. If atoms are analogized to dusty motes, as they are in classical Epicureanism, and "blowne around by winde," they create "an infinite and eternal disorder." Rather, the proper reifications for atoms are fairies, sentient and vital creatures who know what they are doing, even if it is a source of some disorder for the larger inhabitants of their common material universe.

At the end of *Poems, and Fancies*, Cavendish reveals that her materialism and her fantasy are indeed part of the same universe. She expresses her amazement that "any should laugh or think it ridiculous to heare of Fairies and yet verily believe there are spirits: which spirits can have no description, because no dimension." Rather, fairies "are onely small bodies, not subject to our sense, although it be to our reason. For Nature can as well make small bodies, as great, and thin bodies as well as thicke. . . . And if we grant there may be a substance, although not subject to our sense, then wee must grant, that substance, that substance must have some forme; And why not of man, as anything else? And why not rational soules live in a small body, as well as in a grosse, and in a thin, as in a thicke?"[79]

Clearly, neither atomism nor fairies are meant to be a joke, and neither is their author or creator. Since Cavendish thought that all such speculation was probabilistic and that science could entertain as well as enlighten, she could with all good conscience combine atoms and fairies. The fairies emphasize the vitalistic character of her matter theory, even in its first inception. The cosmogony and beatific vision give her work a metaphysical grounding that both amuses the reader and protects the work—her child, and herself—from the charge of atheism. The prominence of motion in her system allows her to incorporate the ontology of the mechanistic universe into her materialism but also allows her to claim originality for her work. The fusion of fairies and atoms underscores the unity of her work, in which both natural philosophy and fantasy function as complementary products of a unitary mind. The rational matter includes the faculties of reason and imagination, both of which ultimately convey the same understanding of nature. This unity will characterize all of her later works, including *Philosophicall Fancies*, which was published shortly after *Poems, and Fancies*, and *Philosophical and Physical Opinions*, published in 1655, which incorporated most of *Philosophicall Fancies* and gives a less fanciful rendition of her developing natural philosophy.

# chapter three

# The Life of Matter

❧

*W*hen Margaret Cavendish began to write natural philosophy, she joined what was becoming the most important intellectual activity of the day, an enterprise that had already produced the three major mechanical philosophers of the seventeenth century—Gassendi, Descartes, and Hobbes. But writing natural philosophy presented a unique set of challenges. Every thinker who attempted to understand the nature of the physical world also had to consider questions that would (presumably) not trouble future scientists: theological, ethical, epistemological, and political challenges. Questions about the creation necessitated questions about the creator. Any discussion of matter had implications for the understanding of order, disorder, and God's providential plan for the universe. Epistemology had to be explained in terms of the actions of matter and motion. The new worlds revealed by telescopes and microscopes, or discovered in other parts of the planet, raised questions about the parameters of humanity and rationality. A new science of nature meant a new philosophy in general.

In the two philosophic treatises Cavendish wrote in the 1650s, she considered many of these questions, if somewhat unsystematically. Her two goals in *Philosophicall Fancies* (1653) and *Philosophical and Physical Opinions* (1655) were, first, to articulate an unmistakably vitalist theory of matter and, second, to develop a system that was unquestionably original.[1] She achieved her first aim, but not without anxiety about the theological and ontological implications of an infinite universe formed from eternal, self-conscious, self-regulating, and always moving matter. She tried to reconcile an increasingly autonomous matter with a providential God but was often forced into fideistic resignation about the deity. She was defensive about her own orthodoxy, which did not prevent her from developing a materialist kind of religion in *Philosophical and Physical Opinions*.

As to her other aim—originality—one might say that Cavendish's natural philosophy was developed in opposition to the many doctrines about nature advanced during the seventeenth century, including mechanism, alchemy, and medicine.[2] But in opposing them, she also appropriated, perhaps unconsciously, some of their approaches to nature. While she openly repudiated atomism in 1655, she retained the idea of a universe composed of matter in motion. Her philosophy includes three kinds of matter: rational, sensitive, and dull or inanimate, of which at least the first two possess innate motion or life. Cavendish used the insights and language of other philosophies in an attempt to understand both the regularity and irregularity of the physical world. She wanted to come to terms with the problem of change and disorder in a materialistic universe where matter is self-determining and vital.

In her material philosophy, Cavendish empowered rational matter, possessed by all creatures, and made it independent of the constraints of sensitive matter or dull matter. Her philosophy, as it evolved in the 1650s, became increasingly more speculative. The properties of matter itself validated her fancies: "I found a naturall inclination, or motion in my own braine to fancies."[3] The motion of her brain also allowed her to develop a unique natural philosophy. Unlike other material philosophers who attempted to find the constituents of material being in order to explain or construct the world of objects they observed, Cavendish assumed that minute parts of matter constituted both the real and the imaginary, the seen and unseen, and every kind of so-called spirit. Her vision of the material world was broader than that of her contemporaries. She saw and imagined matter in everything, and in her thought, even the imaginary became concrete.

Cavendish's natural philosophy is often described as vitalistic materialism, which requires some degree of explanation. Whether matter was passive or active, inert or moving, dead or alive preoccupied many natural philosophers both before and after Cavendish. In the decades before Cavendish wrote, neo-Aristotelians, atomists, and alchemists incorporated living principles into their philosophies. John Henry, Margaret Osler, and Antonio Clericuzio, among other revisionist historians, have argued that even the most mechanical of philosophers, such as Pierre Gassendi and Robert Boyle, integrated some degree of hylozoism and vitalism into their concepts of matter.[4] So in her own version of atomistic or corpuscular matter theory, Cavendish joined many other contemporary thinkers, some of whose ideas are just as strange to modern sensibilities as are Cavendish's atoms and fairies. Her own self-defined concept of living matter, which would change over time, included the idea that matter is agile, self-moving, self-conscious, sensate, and powerful. In *Philosophicall Fancies*, Cavendish wrote, "Life is the Extract, or spirit of common matter; this extract is agile, being always in motion. . . . This essence, or life, which are spirits of sense, move of themselves." She argued that matter "quickens" when it is sufficiently thin; quickening is the same as motion, and motion is the same as life.[5] Cavendish was careful to explain that while she refers to "spirits," she does not mean to suggest that there is any incorporeal element in her materialism.

Although Cavendish was inconsistent in her usage, she often referred to matter as innate or innated matter. At the beginning of *Opinions*, where she states her reasons for rejecting traditional atomism, she writes, "Every atome must be of a living substance, that is innate matter, for else they could not move, but would be an infinite dull and unmoving body."[6] Internalized motion is an inherent quality of some kinds of matter, and since Cavendish believed that matter is eternal, it obviates the necessity of either God or immaterial spirits animating or infusing motion and activity into matter. This concept of living matter separates Cavendish from the atomists like Pierre Gassendi who argued that motion was infused by God into matter, or from the alchemists and Neoplatonists who identified some kind of immaterial spirit as the formative or seminal principle of nature, although her material philosophy borrowed analogies from both. Her ideas perhaps come closest to the neo-Aristotelians and natural magicians who attributed real qualities to matter, such as sympathy and antipathy.[7]

Epistemological and perceptual questions related to her materialism prompted Cavendish to develop a theory of knowledge that depended on the different functions of the three different kinds of matter she envisioned to explain how beings know. Postulating that everything made out of matter possesses some degree of rational matter, Cavendish argued that every kind of material being understands the universe in its own way. In an epistemological sense, all beings are equal.

The homogeneity of matter suggested other problems. Continuously moving matter, while sometimes dancing, was always pushing. Politics in both its philosophic and practical manifestations was never far from Cavendish's thought. Probably inspired by Hobbes, she understood the constituents of both the natural and human worlds as equally driven by the desire for power. She understood disease in the body as analogous to upheaval in the state. She viewed traditional hierarchy as the answer to anarchy in nature and society, while conceding the stabilizing effect of tyranny.

When Cavendish thought about the universe, she considered not only what she knew but also what she could suppose. Both *Philosophicall Fancies* and *Opinions* include musings about the possibilities of other worlds, inhabited by other kinds of beings. Her desire to be considered a very serious natural philosopher never precluded Cavendish's flights of fancy. But these works show that she was fashioning the role that in the next decade would lead her to the Royal Society: a woman natural philosopher.

## PHILOSOPHICALL FANCIES

Cavendish found that she had not exhausted her topic in *Poems, and Fancies* and immediately wrote a follow-up to be published with it. Unlike the first work, *Philosophicall Fancies* was a description of her scientific ideas largely in prose rather than in verse, meant to be taken less as entertainment and more as a serious scientific treatise. Unfortunately, she got the manuscript to the press a week late, and so the volume appeared separately from *Poems, and Fancies*.[8] In this work, she returned to the topics of matter and motion, and life and death. *Philosophicall Fancies* included the first elaboration of the natural philosophy she would develop in later works.[9] It also incorporated alchemical elements into Cavendish's materialism, showing how her mature vitalism was inspired by her acquaintance with alchemy.

Unlike *Poems, and Fancies*, which began with a cosmogonic poem describing the creation of the universe, in this text Cavendish described matter as infinite and eternal and having control over itself, eliminating the necessity of either Nature or her ministers. Cavendish also made it independent of time: matter lives always in the present as a function of its eternity. Moreover, it is unlimited and takes form only when divided by motion. In this sense, it goes from the smallest to the largest, with the very smallest still sometimes described as an atom. In a sense, matter absorbs the attributes of divinity: it is infinite, eternal, omnipresent, and one.

Motion and figure remained important in this second work, but matter seems to have gained at the expense of motion. And although matter "tends to unity," some matter has power over others forms of matter; it is described as "innate matter." This kind of matter uses its stronger motions to rule what Cavendish calls dull or inanimate matter: "This Innate Matter is a kind of God, or Gods to the dull part of matter, having power to forme it, as it please: and why not every degree of Innate Matter be, as severall Gods, and so a stronger Motion be a God to the weaker, and so have an Infinite and Eternall government? As we will compare Motion to Officers or Magistrates. The Constable rules the Parish, the Mayor the Constable, the King the mayor, and some Higher power the King: Thus infinite powers rule Eternity."[10]

Thus, a hierarchy of matter emerges that is both theological and political. Innate matter is both God and king; presumably the universe is a kind of theocracy. The ruling matter governs the dull matter of the universe, its populace, and its motion functions like the officers and magistrates of a kingdom. This analogy is repeated later in the text: "For those Spirits [a term Cavendish uses interchangeably with matter], or Essences, are the Guiders, Governours, Directors; the Motions are but their Instruments, the Spirits are the Cause, Motion but an effect therefrom: For that thin Matter, which is Spirits can alter the Motion, but Motion cannot alter the Matter, or Nature of those Essences, or Spirits."[11]

Political metaphors informed Cavendish's description of matter in this volume, as they had in *Poems, and Fancies* and will in her later natural philosophy. All is not well in the material world because "the severall Degrees of Matter, Motion, and Figure, strive for Superiority, making Faction [Cavendish's footnote: "Which is Likeness"] and Fraction, by [footnote: "unlikeness"] antipathy."[12] Somehow, this disunity must be resolved so that a harmonious material realm will emerge. It is motion that gives one part of matter power over another, even allowing a smaller to vanquish a larger, just as a mouse can kill an elephant. This analogy recalls Hobbes's state of nature in which all live in fear because "the weakest has strength enough to kill the strongest."[13] While Hobbes's humans created the state to control anarchy, innate matter "governs by degree, /According as the stronger Motions be."[14]

Presumably, innate matter is self-conscious if it rules over other matter, but does that mean it is alive? At first, Cavendish suggested that only concretions of matter live: "Thus the Dispensing of the Matter into particular Figures by an Alteration of Motion, we call Death, and the joyning of Parts to create a Figure, we call Life. Death is a Separation, Life is a Contraction." But, in actuality, life always exists in innate matter; although figures may dissolve, the innate matter that composes them will simply join itself with other dull matter and create a new being. As we saw above, "Life is the *Extract*, or *Spirit of Common* [Footnote: "For when Matter comes to such a degree it quickens,"] *Matter*: This *Extract* is Agile, being alwayes in motion; for the Thinnesse of this *Matter* causes the subtlety of the Quality, or property which Quality, or property is to work upon all *dull Matter*."[15] Its substance never loses motion and life. In *Philosophical and Physical Opinions*, Cavendish wrote, "I have said that the innated matter is the thinnest part of onely matter. . . . The innated matter is the infinite extract of the entity of infinite matter, it is the quintessence of nature."[16]

Apparently the crucial determinant of life is how thin matter is: the thinner, the more agile and sharper its motion will be "to cut and divide all that opposeth their way."[17] Cavendish also used thinness as one of the defining characteristics of fairies in *Poems, and Fancies.* Motion becomes innate in the thinnest form of matter, and, in fact, this kind of matter is self-moving: "This Essence, or Life, which are Spirits of Sense, move of themselves: for the dull part of Matter moves not, but as it is moved thereby."[18] The only matter that moves mechanistically in this account

is dull matter, what Cavendish will refer to in her later natural philosophy as inanimate matter.

But while all innate matter includes a principle of life, this vital matter is itself
composed of two kinds: rational and sensitive matter. Sensitive matter presents
sense impressions to the rational matter, which then uses this information to produce conceptions.

The motion of the rational matter "makes Figures; which Figures are Thoughts,
as Memory, Understanding, Imaginations, or Fancy, and Remembrance, and
Will." Unlike sensitive matter, which works directly on dull matter, the rational
spirits are internalized principles of harmony and order which by "casting and
pleasing themselves into Figures make a Consort, and Harmony by Numbers"[19]
Whether or not this is a conscious reference to a kind of Pythagorean cosmic
dance, it is clear that the motion of the dance has become a prism through which
Cavendish understands the interaction of matter.

In *Poems, and Fancies* the motion of the atoms or the motion of the fairies had
determined understanding and will, in this text, rational matter, also dancing,
produces cognition and imagination. Moreover, the harmony or regularity of
these material spirits is self-directed or free: "Will is to choose a dance, that is to
move as they please, and not as they are perswaded by the sensitive Spirits."[20] The
more rational matter there is in a body, the better its intelligence or judgment:
"But where the greatest number of these, or quantity of their Essences are met,
and joyn'd in the most regular motion, there is the clearest Understanding, the
deepest Judgement, the Perfectest Knowledge, the Finest Fancies, the more
Imagination, the stronger Memory, the Obstinatest Will."[21]

The division of animate matter into two kinds also gave Cavendish the opportunity to develop a theory of the passions and perception. It turns out that the rational spirits govern the internal motion of the body and the sensitive spirits its
external motions. When the rational spirits disagree in their motions, the mind
experiences various emotions. Thus, pleasure is a smooth motion, and pain is an
obstructed motion.[22] Disordered motions in the mind result in passions and af-
fects, although Cavendish declined to specify how this process works in specific
cases: "To expresse those several motions, is onely to be done by guesse, not by
knowledge, as some few will guess at."[23] This remark may be a reference to Hobbes,
who believed pleasure was unimpeded vital motion whereas pain was the opposite. A Hobbesian dogmatism, however, was inimical to Cavendish; clearly she has
not abandoned the epistemological skepticism that characterized her earlier work.

There are, therefore, three different forms of matter—the rational, sensitive, and
dull—possessing different characteristics although made of the same substance:

rational, sensitive, and dull matter. "If it be, as it probably it is, that all sensitive Spirits live in dull Matter; So rationall Spirits live in sensitive Spirits, according to the shape of those Figures that the sensitive Spirits form them."[24] This theory of matter anticipates much of Cavendish's mature materialist vitalism, except perhaps for the ontological status of motion, which while dependent on matter still seems to be differentiated from it. The question of the void is also left open in this discussion. On the one hand, there may not be a void: "In Nature if Degrees may equall be,/All may be full, and no vacuity." On the other, "If Infinite Inequallity doth run,/Then must there be an Infinite Vacuum." An author's note declares, "The Readers may take either opinion."[25] The ontological status of rational matter is also somewhat different; it appears to be almost an adjunct of sensitive matter, which shapes it, but unlike sensitive matter, it may be found only in the mind.

In some places in this text, Cavendish employed alchemical terminology to explicate matter and motion, terms she may have learned from her husband, who was interested in alchemy—or "alchymy," as she called it—or from Walter Charleton, who had translated three treatises by the Paracelsian alchemist Joan Baptista Van Helmont in 1650.[26] The sensitive spirits use vitriol or acid to "cut and divide all that opposeth their way."[27] Both sensitive matter and rational matter are analogized to quicksilver, with which they share a fluid and spherical body and the capacity to form an infinite variety of shapes: "Imagine the rational Essence, or Spirits, like little sphericall Bodies of Quick-silver severall ways placing themselves in severall Figures, and sometimes moving in measure, and in order, sometimes out of order: this Quick-silver to be the Minde, and their severall postures made by Motion, the Passions and Affections."[28]

Quicksilver, or mercury, is one of the three principles of matter in Paracelsian philosophy. Mercury is sometimes used as a simple chemical substance, but it is also sometimes viewed as "the soul of matter, the spirit of life."[29] Cavendish identified rational matter as "the Essence of Spirits; as the Spirit of Spirits. This is the Minde, or Soul of Animalls."[30] The influence of alchemical conceits, as well as of the mechanistic concepts of matter in motion, on her concept of nature is revealed in her text: "Animall Spirits are stronger . . . being of an higher extract (as I may say) in the Chymestry of Nature, which makes the different degrees in knowledge, by the difference in strengths and finesse, or subtlety of matter."[31]

We have met the soul of animals already in *Poems, and Fancies*, where it is the principle of life in atoms and fairies. In Van Helmont, the vital force that animates the seeds of matter is called the *archeus*, which Allen Debus describes as "an inherent part of all seeds be they animal, vegetable, or mineral."[32] Although Cavendish did not follow Van Helmont's concept of the *archeus*—a divine agent united to inert matter in a "psychosomatic unit, in which the psychoid part is merely one aspect of an object, which is inseparably bound up with its matter, that is, its material aspect"[33]—her vitalism does reflect his idea that the *archeus* brings "a subtle material breath of life" to the object or monad.[34] The identification of the vital spirit with the *archeus* is endorsed by Charleton in "Prolegemena," in Van Helmont's *Ternary of Paradoxes* (1650).[35] Van Helmont and Charleton never equated the *archeus* with a form of rational or sensitive matter, and the alchemical doctrine of matter is vastly different from Cavendish's, but nevertheless it is very possible that Cavendish's exposure to these ideas affected her materialism.[36]

In the "Prolegemena," Charleton had also examined the question of whether animals possessed language and reason. Although he does not refer explicitly to an animal soul, he does write, "Many wise men, great Scholars, and extreamly tender, in the point of their allegiance to the Church, have thought it no dishonour, to their Creation, nor Dimunation of . . . the transcendent dignity of Humane Nature, to opinion that the Faculty of Discourse, though in greater degree of Obscurity, may be attributed to brute Animals." Indeed, he argued, drawing on arguments both theological and natural, Saint Jerome conversed with a faun, who then became Christian, and bees know their own hives. Hence, "Whether the Power of Ratiocination, be not in common to some Beasts, as well as to man, 'though Imperfectiori modo [imperfectably made] cannot misbecome, though not the Pulpit, yet the Study of the most rigid Divine."[37]

Animals may have sense, Charleton suggested, and Cavendish agreed with the sentiment, even if she had not read his text. She wrote, "So that all matter is moving, or moved, by the movers; if so, all things have sense, because all things have of these spirits in them; and if Sensitive Spirits, why not rational Spirits? . . . Who knows, but Vegetables and Mineralls may have some of these rationall spirits, which is a minde or soule in them, as well as Men? Onely they want that Figure (with such kinde of motion proper thereunto) to expresse knowledge that way. For had Vegetables and Mineralls the same shape, made by such motions, as the sensitive spirits create; then there might be Wooden men, and Iron beasts."[38]

Not only is all of nature alive because it is composed of sensitive matter, but it also may possess understanding and reason because it also contains an element of rational matter. The only thing that distinguishes man, animals, and minerals is shape. And even if the kind of knowledge the other parts of nature possess is different than man's, "yet it is knowledge." Cavendish concluded, "And Vegetables and Minerals may know, / As Man, though like to Trees and Stones they grow."[39]

In Cavendish, speculation soon leads to fancy, and the next seven pages of *Philosophicall Fancies* are devoted to a poem describing the different kinds of men and beasts who are made out of vegetables and minerals. Matter itself does not change, but the shapes it composes can create another kind of world: "Thus may another World though matter still the same. / By changing shapes, change, Humours, properties, and names."[40] Just as *Poems, and Fancies* had ended with a description of fairyland, *Philosophicall Fancies* climaxed in a fantastical description of a different world populated with diverse beings, such as "a woman out of flowers" and "a man out of metals." While this world may be more thoroughly grounded in the substance of her material philosophy than is the realm of Queen Mab, it equally anthropomorphized material beings. Both texts anticipate the conjunction Cavendish made between the text of her most serious natural philosophic work, *Observations* (1666), and its textual companion, the *Blazing World*. For Cavendish, there is a continuum between science and fiction. Both are equally important in elucidating her thought.

And just as Cavendish did not ignore theological concerns in *Poems, and Fancies*, religion also makes an appearance in *Philosophicall Fancies*. Here Cavendish distanced herself from any hint of atheism, not by inserting a cosmogonic poem at the beginning of the text and an ecstatic vision at the end, but in a more conventional way, which also probably demonstrates the influence of Walter Charleton. The book concluded with a poem, "The Diatical Centers," which praises God, and reflects the changes Pierre Gassendi introduced into Epicurean atomism, changes discussed in Charleton's *The Vanity of Atheism*.[41] As we have seen, Gassendi had rehabilitated atomism by making God the creator of the atoms, which he then infused with motion. Likewise, in Cavendish's poem,

> Great God, from Thee all Infinities do flow,
> And by thy power from thence effects do grow.
> Thou order'dst all degree of Matter, just,
> And tis thy will, and pleasure, move it must.[42]

Agency is taken away from matter, and instead God's will rules the universe through motion. This God is by no means the absent landlord of *Poems, and Fancies*; rather, he is a proactive force who orders the world justly, wisely, and well. The fideism that characterized the earlier work is somewhat modified here. Nevertheless, the conclusion of the poem returns to this theological stance. The reason that man seeks to understand God is that God gave man an active mind:

> But thou hast made such Creatures, as Mankind,
> And giv'st them something, which we call a Minde,

Alwaies in Motion, never quiet lyes,
Until the figure of his body dies.[43]

Motion, which had been deified in Cavendish's first work, here plays the role of a divinely inspired human curiosity. The aim of the brain's motion, however, is not communion with God; rather, man wants to know, out of self-love, "if he,/Came from, or shall last to eternity." This effort is futile; the motion of the brain breaks like water on rocks: "But since none knowes the great Creator, must/Man seek no more, but in his goodnesse trust."[44]

Cavendish concluded the philosophic section of *Philosophicall Fancies* with a list of the many topics she could explore more fully: weather and disease, affection and antipathy between species and parts of matter; harmony and disorder in natural and human organization. She was not satisfied with the description of matter and its dispositions she had developed in the pages of her book. Triune matter, whether alive or not, and whether united with motion or not, still presented problems, which her fancy could not adequately solve, in part because it, as a form of moving matter, could not be pinned down. "But Fancy," she explains, "which is the effect of Motion, is as infinite as Motion; which made me despaire of a finall Conclusion of my Booke; which makes my Booke imperfect, and my Fancies unsettled: But that which I have writ, will give my Readers so much Light, as to guesse what my Fancies would have beene at."[45]

Never one to keep anyone guessing, Cavendish published within two years a more systematic philosophic work that explored many of the themes she first addressed in *Poems, and Fancies* and *Philosophicall Fancies*.


### PHILOSOPHICAL AND PHYSICAL OPINIONS

In 1655, Cavendish republished the first seventy-one pages of *Philosophicall Fancies* as the first part of her first professedly serious natural philosophic work, *Philosophical and Physical Opinions*. This text, in turn, was augmented and reprinted in 1663 and in her last work, *The Grounds of Natural Philosophy* (1668). The theory of matter articulated in the 1655 version remained the centerpiece of all of her future philosophic works, including her critiques of other scientific works in *Philosophical Letters* (1664) and *Observations upon Experimental Philosophy* (1666).

*Opinions* begins with Newcastle's defense of Cavendish's originality and her own apologia on the same subject. It includes no less than four "Epistles" to her readers and one dedication, "To the Two Universities." I have discussed the gendered importance of this dedication in Chapter 1, but it bears repeating. She urged learned professors to receive her book

> without a scorn, for the good encouragement of our sex, lest in time we should grow as irrational as idiots, by the dejectednesse of our spirits, through the carelesse neglects, and the despisements of the masculine sex to the effeminate, thinking it impossible we should have either learning or understanding, wit or judgement, as if we had not rational souls as well as men, and we out of a custom of dejectednesse think so too, which makes us quit all industry towards profitable knowledge being imployed onely in looe, and pettie imployments, which takes away not only our abilities towards arts, but higher capacities in speculations.[46]

But at least for Cavendish, the advantages of female lack of education compensate for its disadvantages. "I have heard," she relates in one of the peritextual essays at the beginning of *Opinions*, "that learning spoiles the natural wit, and the fancies, of others, drive the fancies out of our own braines, as enemies to the nature, or at least troublesome guests that fill up all the rooms of the house."[47] It may be that Cavendish was reacting to the imputation that her ideas were stolen from Hobbes, Descartes, or others, which made her repudiate the atomistic fancies of her earlier works. The 1655 treatise contains Cavendish's disavowal of atomism, "A Condemning Treatise of Atomes." It begins with a rejection of inert atoms, whose purposeless motion, if they existed, would create an infinitely chaotic universe: "I cannot think that the substance of infinite matter is onely a body of dust, such as small atoms, and that there is no solidity, but what they make, nor no degrees, but what they compose, nor no change and variety, but as they move, as onely by fleeing about as dust and ashes, that are blown about with winde, which makes me thinke should make such uncertainties, such disproportioned figures, and confused creations, as there would be an infinite and eternal disorder." If the world were composed of senseless, dusty atoms, Cavendish argued in this "Treatise," the result would be disorder; the world we know could not exist: "But surely such wandring and confused figures could never produce such infinite effects; such rare compositions, such various figures, such several kindes, such constant continuance of each kinde, such exact rules, such undissolvable Laws, such fixt decrees, such order, such method, such life, such sense, such faculties, such reason, such knowledge, such power, which makes me condemn the general opinions of atoms."[48]

Most scholars have interpreted this condemnation to mean that Cavendish was repudiating the atomism of *Poems, and Fancies*, but the system articulated in her first book was, strictly speaking, neither Epicurean nor mechanistic. Rather, Cavendish condemned "the general opinion of atoms." Epicurean atoms would not be able to produce the variety of kinds and reasons found in nature; order and law would cease to exist in such a universe. Cavendish's own "particular opinions" of the figures or motions of atoms, "their wars and peace, their sympathies and antipathies, and many the like," avoid these problems: "I have considered that if the onely matter were atoms, and that every atome is of the same degree, and the same quantity, as well as of the same matter; then every atom must be of a living substance, that is innate matter, for else they could not move, but would be infinite dull and immoving body."[49]

The remainder of *Opinions* explains how the innate matter works. The connection between this book and her earlier opinions is confirmed in "An Epistle to the Reader, for my Book of Philosophy," where she discusses her hesitation about having *Poems, and Fancies* translated into Latin because of the difficulty of finding a good translator of poetry: "For this reason I would have turned my Atomes out of verse into prose, and joined it to this book."[50] Apparently there is no substantive contradiction between these works, which she would have willingly combined if they had been composed in the same style.

Thus, when Cavendish concluded in "A Condemning Treatise," "But the old opinions of atoms seems not so clear to my reason, as my own, and absolutely new opinions, which I call my Philosophical Opinions," she is referring not to a difference of philosophy but to a difference of genre. In fact, her new philosophic opinions reflect her first philosophy, except they are presented as opinions, not poems or fancies: in *Poems, and Fancies*, for example, the verse description of the union of the separate principles that Nature employed to create the world was the following:

> Thus severall Figures, severall Motions take,
> And severall Motions, severall Figures make.
> But Figure, Matter, Motion, all is one,
> Can never separate, nor be alone.[51]

In *Opinions*, Cavendish used prose for the same description. "Nature is matter, form, and motion, all these being, as it were but one thing; matter is the body of nature, form is the shape of nature and motion."[52] The message is the same, but the style is different. In yet another "Epistle to My Readers," Cavendish explained again that although she uses different styles of writing, her ideas remain the same: "I must advertise my Readers that though I have writ different ways of one and the same subject, yet not to obstruct, crosse, or contradict; but I have used the freedom, or taken the liberty to draw several works upon one ground, or like as to build several rooms upon one foundation."[53]

Cavendish saw no contradiction between her poetry and prose, but there is a likely reason why she turned from one to the other in 1655. In between the 1653 *Poems, and Fancies* and the 1655 *Opinions,* her old friend Walter Charleton published his detailed defense of Gassendi's rehabilitated Epicurean atomism, *Physiologia Epicuro-Gassendo-Charltoniana* (1654), which sought to integrate Christian theological beliefs into an Epicurean, mechanistic atomism. Although Charleton was unable to eliminate all traces of vitalism from his work, atoms are in the main described as the smallest indivisible bits of matter, existing in a void, and having the properties of "consimilarity of substance," "magnitude" or extension, "figure" or dimension, and "gravity, or weight."[54] Charleton was well aware that atoms had been analogized to dust motes by Lucretius and others, but like Cavendish, he rejected this association:

> Theodoret . . . positively affirms, that Democritus, Metrodorus, and Epicurus, by their exile Principles, Atoms, meant no other but those small pulverized fragments of bodies, which the beams of the Sun, transmitted through lattice Windows, or chinks, make visible in the aer; when according to their genuine sense, one of those dusty granules, nay, the smallest of all things discernable by Linneus, though advantaged by the most exquisite Engyscope, doth consist of Myriads of Myriads of thousands of true Atoms, which are yet corporeal and possess a determinate extension.[55]

Cavendish had rejected dusty atoms because they were not alive; Charleton rejected them because they were not small enough. But without an energizing internal principle of motion, Charleton needed a way to explain the order of the world. At the same time, he faced the same difficulty that had challenged his intellectual mentor, Gassendi: how to rescue atomism from the charge of atheism. Following Gassendi, Charleton rehabilitated atomism by positing God as the source of the creation and the motion of the atoms. He argued, "(1) That Atoms were produced *ex nihilo,* or created by God, as the sufficient Materials of the World, in that part of Eternity, which seemed opportune to his infinite Wisdom; (2) that, at their Creation, God invigorated or impraegnated them with an Internal Energy, or Faculty motive, which may be conceived the first Cause of all Natural Actions, or Motions (for they are indistinguishable) performed in the World."[56]

Charleton's God is voluntarist and omnipotent.[57] The universe he creates is an adjunct of his will. Atoms are passive until God makes them pregnant with motion. Nature has no part in his system either as an independent creative agent, as she had been in Cavendish's first work, or as an internal vivifying principle, as she becomes in *Opinions*. In the latter work, Cavendish postulated that motion, form, and matter are integrated but that Nature herself—also sometimes referred to as itself—is the totality of matter, which is immanently self-ordering. Nature, motion, and life are all united in one material substance:

> The innated matter is the soul of Nature.
> The dull part of Matter, the Body.
> The infinite figures, are the infinite form of Nature.
> And the several motions are the several actions of nature.[58]

Charleton had solved the problem of the motion and action of the atoms by positing a force outside them; Cavendish solved it by making atoms or matter the source of its own motion and action, as well as endowing these particles with eternity and infinitude. The fact that the internalized principle was female in an earlier incarnation is perhaps a comment on the implicit sexual dichotomy of Gassendi's and Charleton's atomism.

Vitalistic materialism might have satisfied Cavendish's philosophic desires, but she knew that, theologically speaking, she was preaching a kind of heresy that seemed to reduce the action of God to the noninterventionist passivity of the Epicurean gods. Voluntarist explanations of atomism had the advantage of maintaining God's active presence in the world. Cavendish's theory almost made God a *dieu manque*. Her defensiveness about her ideas is clearly indicated in another "Epistle," which immediately precedes her condemnation of Epicurean atomism: "I desire my Readers to give me the same privilege to discourse in natural Philosophy, as Scholars have in schooles, which I have heard speak freely, and boldly, without being condemned for Atheisme; for they speak as natural Philosophers, not as Divines; and since it is natural Philosophy, and not Theologie, I treat on, pray account me not an Atheist, but beleeve as I do in God Almighty."[59]

Seen in the context of this pronouncement of faith, "A Condemning Treatise of Atomes," which immediately follows, comes as no surprise. Clearly, Cavendish was more sensitive to the heterodoxy of her work here than she was in her first book, perhaps because Charleton's work highlighted the unorthodoxy of her scientific ideas. Since she had abandoned the poetry and fancy of the earlier works and wished to be taken more seriously by academics and philosophers in *Opinions*, she may have felt obliged to present a specific defense of her right to speculate in natural philosophy.

And so the very last section of the prefatory material in *Philosophical and Physical Opinions* is entitled "The Text to my Natural Sermon." Cavendish began, "I as

the preacher of nature, do take my text out of natural observance, and contemplation, I begin with the first chapter, which is the onely, and infinite matter, and
conclude in the last, which is eternity." A catechism of her natural philosophy follows, which by now should be familiar to the reader:

> The first cause is matter.
> The second is Motion.
> The third is figure
> Which produceth all natural effects.
> Nature is matter, form, and motion, all these being, as it were but one thing;
>     matter is the body of nature, form is the shape of nature and motion.
> The spirits of nature, which is the life of matter, and the several motions are
>     the several actions of nature.
> The several figures are the several postures of nature, and the several parts, the
>     several members of nature.[60]

In *Opinions*, a fused nature and life becomes the anchor of Cavendish's vitalistic materialism. Matter possesses an internalized spirit or "life," what Cavendish
had called innate matter in her earlier work, which acts through motion and
shape. Nature is no longer a creator being but rather the sum total of vitalized
matter. In her later works, Cavendish continued to employ both meanings of Na-
ture/nature, both a personified force and an internalized aspect of matter. Indeed, in her 1656 memoir, *A True Relation*, she analogized her own writing to the
internalized self-ordering of a creator Nature, writing, "I am industrious to Gain
so much of Nature's favour, as to enable me to do some Work, wherein I may leave
my Idea, or live in an idea, or my Idea may Live in Many Brains; for then I shall
live as Nature Lives amongst her Creatures, which onely Lives in her Works, and
not otherwise Known but by her Works."[61]

Nature is absorbed into her own creation, which takes on a life of its own. Matter lives, unlike the inert atoms of Epicurus and the mechanical philosophers. The
more Cavendish developed her vitalist materialism in later works, the more she
wanted to dissociate herself from the vocabulary of atomism. So in the 1663 edition
of *Philosophical and Physical Opinions*, she proclaimed, "[Since 1653] I have
thought more on it [atomism], and could give Better Reasons concerning Atoms
than I could then . . . ; but now I Wave the Old Opinion of Atoms, for it is not
probable, they could be the Cause of such Effects as are in Nature."[62]

Instead, innate matter will be the principle of nature whose motion produces
sickness and health, animals, vegetables, minerals, planets and stars, the elements,
the passions, and reason. Innate matter is life itself: "For infinite motions are the

infinite life, of the infinite and eternal life, which life, is as eternal matter, being part of the matter it self, and the manner of moving is but the several actions of life; for it is not an absence of life when the figure dissolves, but an alteration of life, that is, the matter ceaseth not from moving, for every part hath life in it."[63]

There is no more ambiguity in Cavendish's philosophy about when life begins. It is always a principle of eternally moving matter. Even dissolution does not destroy the ultimate vitality of matter, which is simply an alteration in the figure that precludes death. Cavendish argued, "There can be nothing lost in nature, Although there be infinite changes, and their changes never repeated." Even if figures dissolve into "dust, as small as Atoms," the matter they are made out of continues to live.[64]

While Cavendish turned to alchemy in *Philosophicall Fancies* to find a vocabulary with which to describe her materialism, *Opinions* is informed by medical terminology. It is the first of her works to include large sections devoted to health, a preoccupation that will continue in the 1663 edition of this text and in *Observations* and *Grounds of Natural Philosophy*. Women were supposed to know something about healing in early modern Europe, and Cavendish's discussion of matter, motion, and figure reflects some knowledge of medical literature.

Thus, even though motion is an eternal property of matter, it possesses certain qualitative differences in its actions on figures and even within matter itself, reflecting medical distinctions. Cavendish had identified four kinds of motion in *Philosophicall Fancies*: attractive, retentive, digestive, and expulsive. In *Opinions* she expanded the list to include contractive and dilative motions. Several of these terms have medical referents: in premodern medicine, the retentive faculty of bodies is the ability to keep food in the body; the explosive faculty refers to drugs that expel noxious substances from the body; digestion, a word also used to describe an alchemical process, characterizes the absorbing of food by the body; and dilation refers to the diffusion of food in the body.[65] In Cavendish's redefinition, a retentive motion is to "fix . . . the matter to one place, as if one should stick or glue parts together." Dilative motion, in contrast, is expansive "as to spend, or extend, striving for space or compass." Explosive motion "shuns all unity, it strives against solidity . . . [and] it disperses everything it hath power on." Digestive motions are the opposite of explosive, they "are the creating motions, carrying about parts to parts, and fitting and matching, and joyning parts together, mixing and tempering the matter for proper uses."[66]

These various motions result in what Cavendish called the "transmigration" of matter, as one figure dissolves into its parts and is reconstituted into another figure. Once again, medical terminology guides Cavendish's thought: the creation of new forms out of old is analogous to "the nourishing food that is received into the stomack transmigrated into Chylus, Chylus into blood, blood into flesh, flesh into fat, and some of the chylus transmigrated into humors, as Choler, Flegme, and melancholy."[67] New forms result from the transmigration of moving matter, but these are all essentially new creations. Cavendish expressly repudiated what she calls the "metamorphizing" of the alchemists, who believed that through their "magick" they could transform one substance into another.[68]

In the section of her book devoted to health, "Of the Motion of The Bodie," Cavendish's incorporation of physics and medicine becomes even clearer, while the integration of matter and motion becomes more problematic. Motion functions both as one with matter and disruptive to matter. "Physitians should study the motions of the body," she advised, "as natural Philosophers, study the motions of the heavens, for several diseases have several motions, and if they were well watched and weighed, and observed, they might easily be found out severally." Observation would reveal what motions create ill health and discombobulate the humors. Colic and agues and fevers could all be understood, "as to make animals, though not live eternally, yet very long."[69]

Motion can produce both health and disease; it can create figures and dissolve them. Cavendish viewed the various motions as engaged in a kind of struggle: "Motion doth not only divide matter infinite, but disturb matter infinite: for self-motion striving and struggling with self-motion, puts itself to pain; and of all kinde of motions the animal motions disturbs most, being most busie, as making wars and divisions, not onely animal figures, against animal figures, but each figure in itself."[70] Motions war against each other, even within the figures they comprise, as each motion tries to be preeminent. The motion of one body assaults the motion of another, while within matter, some motions struggle for preeminence over others. The health of the body, like the health of the state, is ravaged by war, both foreign and domestic.

Political analogy becomes increasingly important in Cavendish's work; the English Civil War informed her perception of matter and its disposition with a sense of danger to the health of the commonweal. In the first part of *Opinions*, she analogized matter in motion to the activities of the members of the state: "There are millions of several motions which agree to the making of each figure, and millions of several motions are knit together; for the general motion of that are figure, as if every figure had a Common-Weale of several Motions working to the subsistence of the figure."[71] Some motions are magistrates, and others are soldiers and merchants. Motions are like craftsmen and farmers; some are like bakers and painters. Altogether their harmony constitutes a figure, just as the citizens of a state compose a commonweal.

But states are often beset by conflict, just as animals often suffer illness. For both, the cause of war is the pursuit of power: "Thus moving matter running perpetually towards absolute power, makes a perpetual war, for infinite, and onely matter is always at strife for absolute power, for matter would have power over infinite, and infinite would have over matter, and eternity would have power over both."[72] Innate matter seeks ontological supremacy over space and time, which in their infinitude challenge the autonomy and self-knowledge of every individuated figure of matter. The result of continuous strife, however, is not anarchy, "For there this is a natural order, and discipline is in nature as much as cruel Tyrannie; for there is a naturall order, and discipline often-times in cruel Tyranny."[73]

In fact, motion itself is a kind of tyrant as it tries to impress itself on figure, but since motion is also a part of matter and figure, "the onely and infinite matter is a tyrant to its self, or rather, I may say, infinite, is a tyrant to motion, and motion to figure, and eternity to all."[74] Although this passage is not exactly the clearest political statement Cavendish ever made, it indicates a tension in her thought that continues in her later work. How can disorder or ill health enter into and be reconciled with a universe—or a body or a state—that is composed of infinite homogeneous living matter? Internal tensions within innate matter may cause discord, but ultimately the fusion of the material, form, and motion of matter should obviate conflict.

Cavendish was as perceptive as Thomas Hobbes in recognizing the implications of materialism. In most cases she resolved the disorder inherent in the universe and the state by endorsing a hierarchy of matter, in which innate matter rules just as a monarch governs the state. But in this text, matter is a tyrant striving for absolute power, who brings discipline to a commonwealth disordered by the motions of its citizens. *Opinions* was published two years after Oliver Cromwell became Lord Protector, bringing his own form of order and discipline to the state, an order Cavendish had observed personally during her trip to England.

One of the reasons for the ambition of innate matter is its frustration in being limited in knowing the infinite and eternal world it comprises. *Opinions* also contains the first detailed articulation of Cavendish's epistemology. To some extent, this discussion is an adjunct to her writings on health. Order and disorder can characterize the brain of an individual; a quiet mind is well regulated and harmonious and possesses "gentle imaginations, a clear understanding, a solid judgment, elevated fancies, and ready memory." As always, reason and fancy are associated, in this case as part of the same organ. However, the brain is not exempt from the irregular motions, "but when this rational innated matter moves disorderly, there arises extravagant fancies, false reasons, misunderstandings, and the like."[75]

Rational matter receives the information it works on through the mediation of sensitive matter. In a theory of knowledge Cavendish called "printing" or "patterning," she argued that sensitive matter makes openings in the five senses "wherein the animal receives light, sound, scent, tast, and touch." Once these perceptions are impressed on the senses, the sensitive matter conveys prints of the object to the rational matter, which then can use it in thought, imagination, memory, or dream. Cavendish drew on art to convey her meaning: the prints or patterns produced by the sensitive matter are like two-dimensional paintings, while the rational figures wrought by the innate matter are like sculptures. The innate matter is free to compose what figures it wants out of this information, but the sensitive matter must reflect the dull matter with which it is immediately connected.[76]

Cavendish's epistemology was perhaps closest to Aristotle's, who argued that the sense faculties were imprinted with some kind of impression or species (remember Cavendish's use of this term in *Poems, and Fancies*), which duplicated everything about the object except its matter. This sensation in turn is conveyed to the intellect, which is able to produce formal knowledge of universals.[77] Aristotle's intellect, however, is not material: it is the soul. The only materialist epistemology available to Cavendish was that of Thomas Hobbes, and she did everything she could to distance herself from his thinking.

Having developed a theory of knowledge, Cavendish argued that different sorts of beings could to some degree know the world according to their own capacities. In *Opinions*, Cavendish expanded the theme of epistemological equality, which she had emphasized in *Philosophicall Fancies*. All beings made of matter (which means all beings) possess knowledge, and these different kinds of knowledge are equal, "For as I said before, there is onely different knowledge belonging to every kinde, as to Animal kinde, Vegetable Kinde; and infinite more . . . : as for example, Man may have a different knowledge from beasts, birds, fish, worms, and the like, and yet be no wiser, or knowing then they."[78] Possibly man might be no wiser than women either; Cavendish's doctrine of the equivalence of the knowledge possessed by all different kinds of beings implicitly endorses the possibility of female equality. Women's knowledge may be the knowledge of their kind, both what they know and how they know.

In *Opinions*, Cavendish never explored the gendered possibilities of her epistemology. But she did explore the possibility of hybrid knowledge. "If there be mixe[d] sorts of creatures, as partly man, and partly beast, partly man, and partly fish, or partly beast, and partly fish," she argues, "yet they are particular sorts, and different knowledges, belonging to those sorts."[79] Whether the infinite matter makes such hermaphroditical kinds is a possibility Cavendish will play with in all her later works. If she can imagine such things, in other worlds at least, perhaps they exist: "As the Sun differs from the earth and the rest of the planets, and earth differs from the seas, and seas from the airy skie, so other worlds differ from this world, and the creatures therein, by different degrees of innate matter. . . . So may worlds differ for all we know." "Who knows," asks Cavendish, using the same phrase she employed to begin her speculations about atoms and fairies.[80]

In *Opinions*, Cavendish compared herself to an explorer. Arguing for the possibility of knowing something more about nature than what was known, she noted the parallel between geographic knowledge and natural philosophy. Before the voyages of discovery, people thought they knew all there was to know about the world, and since the time of Sir Francis Drake and Sir Thomas Cavendish (a collateral relative of her husband's), they think there are no more places to be found. Yet, Cavendish maintained, "there are several parts of the world discovered, yet it is most likely not all, nor may be never shall be."[81] Drake's circumnavigation around the world had touched Cavendish's imagination. In her *Poems, and Fancies*, Cavendish imagined Drake as an interstellar explorer:

> There may be many Worlds like Circles round,
> In after Ages more Worlds may be found.
> If we into each Circle can but slip,
> By Art of Navigation in a Ship;
> This World compar'd to some, may be but small:
> No doubt but Nature made degrees of all.
> If so, then Drake had never gone so quick
> About the Largest Circle in one Ship.
> For some may be so big, as none can swim,
> Had they the life of old Methusalem.
> Or had they lives to number with each day,
> They would want time to compasse halfe the way.
> But if that Drake had liv'd in Venus Star,
> His Journey shorter might have been by farre.[82]

Nevertheless, even though perfect and complete knowledge of the globe or the universe is not possible, since we can only know its parts, some knowledge is possible, just as we can gain some knowledge of infinite nature.[83] Thus, the discoveries of new worlds, whether mental or physical, are a search after what may be real.

This modification of Cavendish's original skepticism is an indication that she was becoming aware that if she wanted to be treated as a serious natural philosopher, she would have to present her ideas in a more acceptable style and claim more for their validity. She increasingly believed that her natural philosophy described the real composition of nature, which, while remaining elusive, could in part still be understood. As Cavendish defined her theory of innate matter in more and more detail, she embraced the idea that sense amplified by reason can lead to some degree of knowledge of the natural world, as the observer can reason from effect to cause.[84] The prose works of the mid-1650s show a natural philosopher wrestling with the details of her system and its implications. Between 1653 and 1655, Cavendish expanded her understanding of the nature of matter, which evolved from some kind of atomism to a triune description of matter as rational, sensitive, and inanimate, with the first two kinds defined as innate or vital matter. Motion, which had been an externalized force in *Poems, and Fancies*, becomes increasingly inherent in matter in Cavendish's next two works, although it is sometimes characterized as somewhat disruptive to the internal composition of matter itself.

Cavendish's matter theory led her to theological speculation and a defense of her own brand of orthodoxy. As eternal and infinite matter increasingly preempted the role of any sort of deity in her work, Cavendish was forced to adopt qualifiers in her discussion of God and matter. God has power "to order that moving matter, as that Deity pleaseth," and "though nature is infinite matter, motion and figure creating all things out of its self, for of matter they are made, and by motion they are formed into several and particular figures, yet this Deity orders and disposes of all natures works."[85]

Whether this explanation of divine providence would have satisfied anyone examining Cavendish's conscience in the mid-seventeenth century is certainly debatable. Her belief in the continual strife of matter undermined the idea of a providential ordering force, as it raised the specter of constant disorder in the state. Cavendish's belief that all creatures had epistemological parity might also raise eyebrows: it implicitly suggested that all animals have souls and presumably could be saved. The argument that matter is eternal and infinite also contradicts the Christian doctrines of creation in time. Moreover, Cavendish was more than happy to entertain the notion of the existence of other worlds and other kinds of beings, all the product of ubiquitous matter.

The notion of otherness, of things being different from what she knew, intrigued Cavendish and led her into yet another voyage of fantasy. Cavendish did not yet want to abandon completely a wider and more imaginative exploration of natural philosophy in favor of philosophic treatises. She also sought a wider audience than philosophers. Her next work, *Natures Pictures* (1656), allowed her to throw off the constraints of philosophy and to employ her fecund imagination in the creation of multiple worlds. But at the same time, fancy also became a vehicle to find explanations for those aspects of her philosophy undiscoverable by rational means. Since Cavendish believed that imagination was particularly accessible to women, *Natures Pictures* provided an avenue for speculation—and, as we will see, self-promotion.

# chapter four

# The Imaginative Universe
# of *Natures Pictures*

*I*n *Philosophicall Fancies* and *Philosophical and Physical Opinions*, Margaret Cavendish began her search for a new natural philosophy in earnest. The modification, transformation, and elaboration of the atomism of *Poems, and Fancies* occupied most of her attention and were to a certain extent motivated by her change of genre from poetry to prose. But her distancing from the atomic ontology of matter in motion, and the development of her vitalistic materialism, affected how Cavendish thought about the divine and the natural. In finding a place for God in her system, beyond her early work's vision of a largely absent proprietorship, Cavendish was forced to reconsider the role of nature in her cosmos, either as an external force or as an internal ordering principle of the world. The speculation about the role of nature, either as an innate material principle or as an anthropomorphized figure, continues in *Natures Pictures* (1656).

The result was a new cosmogony in which the role of a gendered nature was interpreted both positively and negatively, perhaps reflecting Cavendish's ambiguous feelings about (most) women, as well as her increased sensitivity to the implications of describing matter as alive. Cavendish sought for the origin of disorder and sin in her cosmos and sometimes found it in female Nature.

*Natures Pictures* is a collection of many imaginative stories with a purportedly moral purpose, providing Cavendish with a forum for exploring diverse topics. It repeats many of the themes and arguments of *Poems, and Fancies* and *Philosophical and Physical Opinions*, but in a forum allowing the untrammeled exercise of her imagination. We hear about her three forms of matter again, but now within the context of a divine marriage or a saint's tale. *Natures Pictures* is also the first of Cavendish's works to reflect an acquaintance with the psychology and moral philosophy of Thomas Hobbes, unlike earlier works that disclaimed any knowledge of his philosophy. Cavendish's discussion of the nature and freedom of the soul and the development of her theories of perception are influenced by Hobbes.

Her own psychology allowed her to argue for the reality of the fantastic. The possibility of the fantastic, in turn, led to stories about other worlds and their possible inhabitants. Clearly influenced by the travel literature of her time and the imaginary journeys of earlier thinkers to other worlds, Cavendish explored what might be over and under the earth and in places not yet discovered by Europeans.[1]

From the publication of the first edition of *Philosophical and Physical Opinions* in 1655 until the last edition of this text in 1668 as *Grounds of Natural Philosophy*, Cavendish separated her philosophic and fanciful texts, although *Observations upon Experimental Philosophy* is bound together with *The New Blazing World*. But the cross-fertilization of her ideas continued, with the fantastical works underscoring and promoting her speculation in natural philosophy, while at the same time letting her imagination roam freely. This tendency is nowhere more evident than in *Natures Pictures*, where any number of genres and literary forms are mixed together to produce a kind of moral philosophy complementing her natural philosophy. Comedies, tragedies, poetry, romances, philosophy, and histories all are meant to teach ethics: "The design of these my feigned stories is to present virtue, the Muses leading her, and the Graces attending her. Likewise, to defend Innocency, to help the distressd, and lament the unfortunate. Also, to shew that Vice is seldome crown'd with good Fortune; and in these Designs or Pieces I have described many sorts of Passions, Humours, Behaviours, Actions, Accidents, Misfortunes, Governments, Laws, Customes, Peace, Wars, Climates, Situations, Arts and Sciences."[2]

For a woman, of course, the exposition of moral philosophy was as problematic as the discussion of natural philosophy or theology. The earl defended his wife's right to write on all these topics by comparing her favorably to Homer, Aristotle, Hippocrates, Cicero, Virgil, Horace, Plutarch, Aesop, Plautus, and the Apocrypha. The range of Cavendish's compositions is indicated by the range of literary company she keeps, and it implies her familiarity with the ancients. As he had before, in a prefatory poem, Newcastle charged that detractors were responding to her sex, not her work:

O but a Woman writes them, she doth strive
T'intrench too much on Man's Prerogative;
Then that's the crime her learned Fame pulls down;
If you be Scholars, she's too of the Gown:
Therefore be civil to her, think it fit
She should not be condemn'd, 'cause she's a wit.[3]

Whether Cavendish is witty or not remains to be seen. As usual, she denigrated her abilities while emphasizing her originality. In "An Epistle to My Readers," she wrote, "Perchance my feigned stories are not so lively described as they might have been," since unlike Van Dyck, who had models for his pictures, hers are not copied "from true Originalls."[4] Nevertheless, Cavendish commended her work, and in particular, she instructed her readers, "Yet I do recommend two as the most solid and edifying, which are named the Anchoret, and the Experienced Traveller, but especially the she Anchoret."[5] Cavendish had originally planned to put parts of "The She-Anchoret" in *Philosophical and Physical Opinions*, and here she suggests that readers compare "those parts or places which treat of the Rational and Sensitive Spirits" with the ideas contained in the earlier composition.[6]

All of Cavendish's wide-ranging speculations in *Natures Pictures* reflect the premises of her natural philosophy, although they sometimes lack lucidity and coherence. Some parts of her work contradict others; some flights of fancy undermine the cogency of her argument. But since Cavendish thought that reason and fancy existed along a continuum of conception and imagination, she felt entitled, and sometimes required, to explore the implications of all her ideas in all possible genres. Clearly she knew she was constructing fantastical stories in this text, but these tales become yet another way of conveying her philosophy. In this chapter, we will see how extraordinary some of her conceptions become.

## "THE SHE-ANCHORET"

Cavendish expressed her developing ideas about God, nature, and the world in "The She-Anchoret," a story about a fictionalized other self published in *Natures Pictures* (1656). The figure of the She-Anchoret is disassociated from other women by her chastity, just as Cavendish was separated from the rest of her sex by her writing. The She-Anchoret's power is in her singularity, and the fame she gains is the projected realization of her creator's ambitions. Her discourse encapsulates Cavendish's natural philosophy and treats its implications for theology and especially ethics. In addition to reimagining nature, "The She-Anchoret" describes the functions of rational and sensitive matter and develops a theory of self-motion that explains the possibility of free will in a fixed universe, as well as examining the source of disorder and sin. Freedom, determinism, and predestination are all understood in terms of the epistemology of innate rational matter or the soul, which allows for its independence from the sense impressions imprinted by the sensitive spirits.

"The She-Anchoret" is distinct from other natural philosophers' treatises on psychology and imagination because it argued for the objective existence of subjective conceptions, including the existence of the marvelous. In other stories in

*Natures Pictures*, Cavendish described fantastic worlds that might or might not exist. The possibility of other worlds and their inhabitants, constructed in different ways from the innate matter, allowed Cavendish to speculate on the nature of humanity and its connection with other kinds and species.

In the "The She-Anchoret," Cavendish fashioned a feminist tale intended to secure for her the approbation she felt she lacked in the actual world of scholars, schoolmen, and natural philosophers. This story is only one of several in *Natures Pictures* in which an astonishing woman instructs men in proper philosophy, who beg to be enlightened through her wisdom. A similar scenario occurs in another romance in the text, "An Assaulted and Pursued Chastity," which anticipates the more famous *Blazing World*.[7] In all these stories, Cavendish restated her material philosophy while awarding its exposition to an avatar of herself. She awarded power to women even when they might have been powerless: the She-Anchoret is a virgin female orphan pursued by a king, and the virginal heroine of "An Assaulted and Pursued Chastity" is a young shipwrecked lady at risk of being sold by a bawd to the Prince of the "Land of Sensuality." Both characters avoid their fates through the exercise of wit and intelligence. "The She-Anchoret" is essentially an embroidered treatise disguised as a saint's life; it might even be called a gothic treatise. It begins with the death of the heroine's father, who enjoins her not to marry, but "Live Chast and holy, serve the Gods above, / They will protect thee for thy zealous love." It ends with the daughter's suicide to avoid bringing war on her country from the importunities of a married king. Her countrymen "made such outcries and lamentations, and mournings, as if there had been an utter desolation of the whole world . . . they buried her with great solemnity, and intombed her costly, the State setting up her Statue of brasse . . . the Church deified her as a saint."[8]

In this unnamed country, Cavendish's projection gains the fame and everlasting glory her creator desired and had asked for in the dedication to *Philosophical and Physical Opinions*. The proclamation, which begins "To the Two Universities," where she asks for the acceptance of her work "for the good encouragement of our sex," also includes a prognostication of her fate if she does not receive their approval:

> I am very confident I shall finde it no where, neither shall I think I deserve it, if you approve not of me, but if I desserve not Praise, I am sure to receive so much Courtship from this sage society, as to bury me in silence; thus I may have a quiet grave, since not worth a famous memory; but to lie intombed under the dust of a University will be honour enough for me, and more then if

I were worshipped by the vulgar as a Deity . . . and who knows, but after my honourable burial, I may have a glorious resurrection in following ages, since time brings strange and unusual things to passe.[9]

Cavendish may have wanted to be buried in a quiet university grave, patiently awaiting the glory future ages would accord her philosophic work, but she craved another, more rapid entombment in fancy. The She-Anchoret is buried twice: once in a cell and once in a grave. Her chosen fate echoes that of other learned ladies of the Renaissance who, according to Margaret L. King, "withdrew from friendships, from the life of the cities, from public view, to small corners of the world where they worked in solitude: to self-constructed prisons, lined with books—to book-lined cells."[10] Cavendish described her own desire for solitude in her autobiography, which was published with *Natures Pictures*: "I being addicted from my childhood to contemplation rather than conversation, to solitariness rather than society, to melancholy rather than mirth." Nevertheless, she chose sometimes to depart her home, "lest my brain should grow barren, or that the root of my fancies should become insipid. . . . I would not bury myself quite from the sight of the world, I go sometimes abroad."[11]

Cavendish's literary celibate has no need to leave her cell to inspire others. Rather, representatives of all social types and professions come to wait on her and be enlightened.[12] Indeed, "She grew as famous, as Diogenes in his tub, all sorts of people resorted to her, to hear her speak, and not only to hear her speak, but to get knowledge, and to learn wisdom, for she argued rationally; instructed judiciously, admonished prudently, and perswaded piously, applying and directing her discourse according to the severall studies, professions, grandeurs, ages, humors of her auditory."[13]

The She-Anchoret is as cognizant of the capacities of her audience as Cavendish was of her readers. In the first two dialogues, with the Natural Philosophers and the Physicians (which in this case means those who study motion), Cavendish's natural philosophy is restated, although with some variations. The fairies, or their material projection, seem to receive short shrift here, as the power of autonomous moving matter is emphasized at the expense of separate principles of nature. The interlocutors ask the She-Anchoret if the kinds of matter she describes are "little creatures." "No," she responds, "they are not creatures but Creators, which creating brains, may easily understand; and those that cannot conceive have a scarcity thereof."[14] We can anticipate that Nature will play a different role in "The She-Anchoret" than she did in *Poems, and Fancies*. Although eternal matter is the source of its own creation in the vitalistic materialism of *Philosophical and Physi-*

*cal Opinions*, in this text God and Nature are reimagined as the parents of creation: "Then they asked her what Deities she thought there were? She answered, she thought but one, which was the father of all creatures, and nature the mother; he being the life, and nature the only matter, which life and matter produceth motion, and figure, various successions, creations, and dissolutions."[15]

The abbreviated cosmogony depicted here gives a much more corporeal role to a male-figured divine principle that progenerates with a material nature, producing motion and figure. Presumably, matter or Nature becomes vital when God, here referred to as "the Diatticall Life," infuses vitality into her. (The adjective "diattical" recalls Cavendish's fideistic poem at the end of *Philosophicall Fancies and Opinions*—"The Diatical Centers"). God certainly sounds like the active principle, and nature seems passive. This analogy sounds closer to Walter Charleton's hylozoic atomism than does the solitary female creator of her earlier work. In another poem in *Natures Pictures*, the matrimonial state of Nature and Jove, who is an anthropomorphized god, is even more explicit and connected directly to Cavendish's vitalistic materialism:

> But who can tell that Nature is not Wife
> For mighty Jove? and he begets the life
> Of every Creature which she breeds and brings
> Forth sevral Forms, each Figure from her Spring.
> Thus Souls and Bodyes in one Figure joyn,
> Though Bodyes mortal be, the Soul's divine,
> As being begot by Jove, and so
> The purest part of Life is Souls, we know;
> For the innated part from Jove proceeds,
> The grosser part from Natures self he breeds
> And what is more innated than Mankinde,
> Unless his Soul, which is of higher kinde.[16]

This poem, at least, urges Nature's children to respect her, but in "The She-Anchoret," when Nature starts to act independently in creation, she is lazy and incompetent; the She-Anchoret explains Nature's faults by stating, "The most prabable [sic] reason I can give; that nature for the most part works so imperfectly, is that she hath so much work to do, as we may say, she that not so much leisure to be exact."[17] Nature sounds like an overworked housewife—and her image is about to become even more tarnished.

When the She-Anchoret turns to moral philosophy, God triumphs and Nature falls in a theodicy more orthodox than anything we have encountered in

Cavendish's earlier writings. Evil comes from nature and good from God/Jove: "All evil lives in nature as all good in Jove, for in nature, said she, is discord, in Jove concord, by nature confusion, by Jove method; and though, said she, Joves goodness and power will not suffer man to run into confusion, yet nature, said she, struggles and strives like an untoward jade, that would break lose to run wildly about, and the skittish tricks, said she, are the sinnes against Jove."[18]

This characterization of nature as a source of disorder and sin seems to confirm Carolyn Merchant's contention that in the sixteenth and seventeenth centuries, "wild incontrollable nature was associated with the female."[19] Indeed, the image of nature as a skittish jade plays off the notions of both female bestiality and viciousness. This description might be compared with another example of misogyny occurring in one of Cavendish's comic tales in *Natures Pictures*. When a hapless male servant, following the commands of his master, attempts to part two battling common women, "it did so enrage their fury, as they left fighting with each other, and fell upon him, where to help himself he was forced to fight with them both, at last it grew to be a very hot battle." They all end up in the stocks, where the scolding women continue the poor man's humiliation.[20]

The woman as scold is a commonplace of popular culture from Punch and Judy shows to Shakespeare's shrewish Kate. But such behavior was not only a source of humor but also an affront to traditional forms of order. When a woman reprimanded a man, she was turning the world upside down.[21] Unruly women eventually had to be controlled in order to preserve society. In order to preserve the order of the universe, explains the She-Anchoret, Nature had to be controlled by her husband: "The goodnesse and power of Jove, said she, doth still hinder nature from running to confusions, and rectifies the disorders therein: for War lives in nature, said she, and peace in Jove."[22] God's rectifying influence is particularly important because Nature not only introduces disorder into the universe; she also is the cause of sin. She is Eve, corrupt and corrupting. All of nature shares her sin; the original sin of humankind becomes the property of the natural world.

But the discourse of the She-Anchoret may be read in a different way. The identity of "nature" shifts in the discussion of moral philosophy from an anthropomorphized principle to an internalized principle of vital matter. This change reflects the similar fusion of matter and nature in *Philosophical and Physical Opinions*, described in the previous chapter. In that case, the root cause of this union may have been Cavendish's uneasiness about traditional atomism's heretical implications. In this case, Cavendish also exhibited her increased sensitivity to religious objections to her work, particularly as she redefined her materialism. "Nature" migrates in the course of this passage from "she" to "all things in nature" to "itself ": "And as nature is apt, said she, to commit sins against Jove, so nature is apt to disorder, crosse and vex it self." Disorder in *Poems, and Fancies* was the result either of disputes among the four ministers of Nature or war between the atoms as they jostle one another at Motion's command:

> When Motion, and all Atomes disagree,
> Thunder in Skies, and sicknesse in Men bee.
> Earthquakes, and Windes which makes disorder great,
> This when that Motions all the Atomes beat.[23]

But in "The She-Anchoret," when vitality is internalized in matter and motion, so is the cause of disorder in nature. As we have seen, all matter strives for power, and the result is disorder. The logic of Cavendish's own materialism made it necessary to integrate confusion into her ontology. "All things in nature" are inundated with "excesse, mischief, and cruelty, as to strive to destroy to no use, to obstruct to no purpose, to slander the Creations, to displace Creations, to oppose a right, to defend falsehood, to wrong Innocency, to hurt the helplesse, to destroy the hurtlesse."[24] The theme of a disordered and baroque nature will reappear in *Observations* as an explanation for monsters and wonders and in *Grounds of Natural Philosophy* as a description of a miserable irregular world.[25] Internalized nature, in this case, allows Cavendish to shift the confusion and discord in Nature from a female principle to all creation. "All things," the She-Anchoret explains, "in nature are guilty as much as man in one kind or other."[26]

However, the very worst sins that the creatures in nature commit against Jove are not to believe he is above nature, or to think it is Nature—and not the knowledge and power of Jove—that "governes so wisely, that orders so prudently, that produceth so orderly, that composes so harmoniously, and, all with a Free-will, a pure goodnesse, and infinity bounty."[27]

Clearly this is a preemptive strike against anyone who might have suggested—and we can think of one person—that Nature had a more critical role in creation than God. Was Cavendish repudiating the cosmogony of *Poems, and Fancies* in this text? In other parts of *Natures Pictures*, a personified nature is conceived as a royal lady who lives in splendor and rules with the aid of beings who are the reified reflections of Cavendish's material philosophy:

> The Rational Creatures are her Nobles.
> The Sensitive Creatures are her Gentry
> The Insensible Creatures are her Commons.
> Life is her Gentleman-usher.

Time is her Steward.
And Death is her Treasurer.[28]

In fact, this is the very same Nature whom Cavendish described in her first work and whose actions she wanted to describe here in prose: "I had a Design to put my Opinions of my Atomes in Prose, as thinking Verse not so proper for Philosophy: but finding it would put me to charge of labour and study, and not likely to be well done, I desisted." What follows is a very condensed summary of her cosmogony: "Nature, when she made the World, thought it best to call a Councel; for though she had power to Command, yet there must be those that must execute her Authority. Her Councellors were four, Matter, Form, Motion, and Life."[29]

Matter is described as "grave" and "solide"; Form has a "clear Understanding" but is "unconstant and Facile"; Motion is "subtil, ingenious" with "quick Wit"; and Life is "weak" at the beginning and "dull" at the end but in between "very strong" and possessing "strong reason." The principles thus possess more personality than they did in *Poems, and Fancies* and are far more contentious, with Matter and Form allying together against Motion and Life: "Which two Factions many times disagreeing, their Councels did antipathize; and often crossing and thwarting each other, caused so many Obstructions, and Contradictions, and Imperfections in Natures Works as are which caused great Troubles in Natures Government."[30] In this case, disorder originates with the four principles, not with Nature, who produces neither error nor sin.

The picture of a benign Nature informs other stories in *Natures Pictures* as well, including a description of a man's search for answers about his soul, which is a mirror image of the "The She-Anchoret"—only in this case, scholars, courtiers, soldiers, and chemists try to answer the man's questions and only demonstrate their own ignorance. They are not as wise as Cavendish's projected self, in this story represented by a wise old couple who, after warning about the presumption of inquiring about God and Nature's work, inform their interlocutor:

The blessings which Jove gave unto Mankinde
Are peacefull Thoughts, and a still quiet Minde;
And Jove is pleas'd, that when that we serve his Wife,
Our Mother Nature, with a Virtuous Life;
For Moral Virtues are the Ground whereon
All Jove's Commands and Laws are built upon.[31]

The many different depictions of a feminized Nature in *Natures Pictures* reflect the ambivalence Cavendish felt about herself and her sex. Even in fancy, she could not contemplate the image of an independent female without at least some anxiety, even if that person is saintlike in her wisdom. The fate of the idealized She-Anchoret is not happy. In repulsing the advances of the inflamed king and choosing virginity—an explicit challenge to male control—the philosophic celibate essentially brings disorder and war on her own country, and her only possible recourse is to commit suicide, in its essence a sin against God and nature. Her future deification is self-defeating when viewed against the expectations and morality of her times. She is a saint without sanctity. Her retreat from the world is really an attempt to describe the world, not to abjure it. Godly women in the seventeenth century were expected to be discreet about their piety and its expression. According to Patricia Crawford, the life of prayer was "a secret life."[32] But the She-Anchoret is an exhibitionist. By teaching all the (male) divisions of society, she becomes a kind of monster; she is an example of what Marina Leslie calls "the monstrous usurping androgyne."[33]

Moreover, she is not even a good philosopher because she crosses disciplinary borders. The She-Anchoret advises against the intermixing of religion and science: "As if a Physician [a student of motion] should study Theologie, he will neither be a subtill Divine, or an eloquent Preacher, nor a knowing Physician, likewise those who study Naturall Philosophy and also Theology, one study confounds the other, For Natural Philosophy proves a God, yet it proves no particular Religion."[34] This caution does not stop the She-Anchoret, who perhaps thinks herself exempt from her own prohibitions. Indeed, she expounds on the most complex and controversial subjects of her time, including the nature of pleasure and pain, the debate about free will and predestination, and the question of the rationality of animals. Just as *Natures Pictures* is a medley of different forms and genres, "The She-Anchoret" is a polyphony of divergent subjects and disciplines. Both find unity only in the person of their expositor, whether Margaret Cavendish herself or her saintly incarnation.

## THE IMPLICATIONS OF A MATERIALIZED NATURE

Cavendish might have wanted to make herself into a saint because she feared being associated with the atheism of the archmaterialist Epicurus or with the reputedly atheistic Hobbes. Cavendish had already disposed of Epicurean atomism when she published *Natures Pictures*, although the book does feature a condensed version of her original atomism. It is mentioned as one of a number of natural philosophies by a Learned Lady in an imagined dispute between herself, a Wise Lady, and a Witty Lady.[35] Nevertheless, the concerns that caused Cavendish to announce that she was not an atheist in *Philosophical and Physical Fancies* and prompted the She-Anchoret to articulate a new kind of theodicy in *Natures Pictures* continued to haunt their author. In fact, they may have been exacerbated by the many different genres Cavendish utilized in *Natures Pictures*. Romance, poetry, dialogue, fantasy, and hagiography all reflect in some degree the implications of a materialist and vitalistic natural philosophy, and while Cavendish may have distinguished her thinking from atomism, her worldview still envisioned a particulate material basis of the universe and its inhabitants. Despite her attempt to distance herself from the mechanical philosophy, Cavendish shared many of the ideas of Epicurus and his modern analogue, Thomas Hobbes.

According to contemporaries, the mid-seventeenth century was replete with atheists. Divines and scientists alike pronounced that England swarmed with people denying the existence of God. These alarmists included Walter Charleton, who in 1652 wrote that England had more "atheistic monsters" than any other period or place.[36] But atheism was an umbrella term in the seventeenth century and included many different forms of heterodoxy. Denial of the immortality of the soul and God's providence were also considered atheistic, as were materialism and naturalism.[37] The philosophy of the ancient materialist Epicurus was particularly problematic in this case because of his doctrine that the atoms constituting the soul shared the general dissolution of matter after death. Moreover, Epicurus's material gods did not bother themselves with the care of the universe but were the epitome of the ideal of tranquility in the mind and pleasure or lack of pain in the body that his moral philosophy posited as the highest forms of happiness.[38] While Cavendish's Jove and Nature are much more active than that, her divinities share the materiality of the Epicurean gods: they progenerate.

There are also Epicurean tendencies in Cavendish's moral philosophy. The She-Anchoret echoes Epicurean ethics when she pronounces, "Health and pleasure is a Heaven, which gives the body rest, and the minde Tranquillity."[39] Whether health and pleasure is *the* Heaven is left an open question, but this description is immediately followed by the statement that "Light is the beatificall vision." We have seen this sentiment before, in the ecstatic and pantheistic poem that concluded Cavendish's atomic poems in *Poems, and Fancies*. At least when it comes to her "natural" religion, it seems that Cavendish's notion of god and the soul was material.

This was a troubling conclusion, and Cavendish's work reflected her uneasiness with it. It was too close for comfort to the philosophy of the archmaterialist Hobbes, who was repeatedly charged with atheism and Epicureanism because he denied the existence of incorporeal substance; for Hobbes, both God and the soul

were corporeal. Hobbes had argued in *Leviathan*, "Substance and Body, signifie
the same thing; and therefore Substance incorporeall are words, which when
they are joined together, destroy one another, as if a man should say Incorporeal
Body."[40] The Cambridge Platonist Henry More's response to this claim was axi-
omatic: "No Spirit, no God."[41]

Cavendish often conflated the soul and rational innated matter, the primary
component of matter in her vitalistic materialism. Her doctrine of the soul is emblematic of the use of correspondences in her thought, as well as the power metaphorical physicality had on her thought. The first thing the She-Anchoret explains
to the Natural Philosophers is that there is a tripartite soul in man: "Man hath three
different natures or faculties; a sensitive body, animall spirit, and a Soul, this soul is
a kinde of a Deity in it self, to direct and guide those things that are inferior to it, to
perceive and descry those things that are far above it; and to create by inventions:
and though it hath not an absolute Power over it self, Yet, it is an harmonious &
absolute thing in it self."[42] The soul is the seat of cognition and is structurally
equivalent to God, with whom it shares a providential ordering of lower creatures
or faculties. Although its external workings may be affected by elements outside its
control, its internal workings are fully within its own control. While it has a connection with the body (here described as "sensitive" matter), it is in some sense separate from the body: "And though the sensitive body hath a relation to it, yet no other
ways than Jove's mansion has to Jove, for the body is the only residing place, and the
Animall Spirits are as the Angels of the soul, which are messengers and intelligenc-
ers."[43] Soul is to body as God is to his physical shell; the animal spirits are the inter-
mediaries between the soul and the body, a role the angels play in the divine sphere.

But the relationship between man and God may be more than simply sym-
metrical. In another story in *Natures Pictures*, the soul is described as the child of
Jove, while the body is produced by Nature:

> Thus Souls and Bodyes in one Figure joyn,
> Though Bodyes mortal be, the Soul's divine,
> As being begot by Jove, and so
> The purest part of Life is Souls, we know;
> For the innated part from Jove proceeds,
> The grosser part from Natures self he breeds
> And what is more innated than Mankinde,
> Unless his Soul, which is of higher kinde.[44]

This union of mind and body, produced by the two different creative principles, clearly duplicates the fusion of innate and inanimate matter. While the soul

is "the purest part" and may be "of higher kinde," it is identical with innate matter. A corporeal god begets a material soul.[45] This bit of speculation, however, is preceded by Cavendish's typical equivocation: "Who can tell," she asks, whether Jove really married Nature? Indeed, the speaker says,

Who knows, said he, first Cause of any thing,
Or what the Matter is whence all doth spring?
Or who at first did matter make to move
So wisely, and in order, none can prove.[46]

Cavendish was fully aware that at the time she wrote many people had very definite opinions on all the questions she thought beyond human knowledge. England had just fought a civil war partly to answer these queries. The She-Anchoret was aware of the historical implications of religious diversity: "The opinions men have of Jove are according to their own natures; and not according to the nature of Jove, which makes such various Religions, and such Rigourous Judgment in every Religion, as to condemn all but their owne opinion, which opinions are so many and different, as scarce any two agrees, and every opinion judges all damned, but their own, and most opinions are, that the smallest fault is able to damne, but the most Vertuous life, and innocent thoughts not sufficient to save them."[47]

Like many other moderate Christians of her time, Cavendish feared religious enthusiasm. Religious passion could only disrupt the state, a sentiment Cavendish shared with Thomas Hobbes and the future members of the Royal Society.[48] Her fideistic theology reflected the contemporary politico-religious climate as well as her skepticism. The She-Anchoret warns the Holy Fathers of the Church who come to visit her, "The Preachers for heaven, said she, ought not to preach factions, not to shew their learning, nor to express their wit, but to teach their flock to pray rightly." Authentic prayer, she instructs, is "a zealous flame raised from a holy fire kindled by a spark of grace in the devout heart, which fills the soul with admiration, and astonishment at Joves incomprehensible Deitie."[49] Theology is too dangerous a business for the laity, who should simply worship a God of wonder without seeking to know more about an unknowable being.

Nevertheless, when the scholars who study theology come to hear her wisdom, the She-Anchoret answers their very dangerous questions. No issue was more explosive at this time than whether human beings were determined in their actions while they lived and predetermined to salvation and damnation after they died. A universe composed of matter in motion, with no additional incorporeal principles, was considered particularly dangerous because it implied that all action was the result of the mechanistic impact of particles of matter. This doctrine was particularly associated with Hobbes, who discussed it in *Leviathan* and most specifically in a series of treatises he wrote at the instigation of Cavendish's husband, then Marquis of Newcastle.

At the time that Cavendish was writing *Philosophical and Physical Opinions* and *Natures Pictures*, a letter Hobbes wrote to Newcastle, "Of Liberty and Necessity," was published without the knowledge of either party. It contained a recasting in written form of a debate between Hobbes and John Bramhall, bishop of Derry, which had occurred in 1645 in Newcastle's presence. In 1655, Bramhall replied to this letter, and the next year Hobbes wrote a book-length rebuttal of his critic entitled *The Questions concerning Liberty, Necessity, and Chance*.[50] This lengthy dispute, which continued after the Restoration, focused on the issue of free will and determinism. While Cavendish might not have been familiar with any of the published texts, it is inconceivable that she was not aware of the discussion of this burning topic.

The implicit closeness of her idea of physical determinism to that of Hobbes is revealed in the She-Anchoret's answer to the moral philosophers who ask her what chance and fortune might be: "Chances, said she, are visible effects from hidden causes; and fortune a sufficient cause to produce such an effect, for a conjunction of many sufficient causes to produce such an effect, since the effect could not be produced, did there want any one of those causes, by reason all of them together were to produce, but that one effect."[51] Thus, causes necessarily follow from effects, and chance or fortune is simply ignorance of the causes of effects. Compare this passage with Hobbes on the same subject: "That which I say necessitateth and determinateth every action, that my Lordship [Newcastle] may no longer doubt of my meaning, is the sum of all things, which being now existent, conduce and concur to the production of that action hereafter, whereof if any one thing now were wanting, the effect could not be produced."[52] And Hobbes concluded, "This concourse of causes, whereof every one is determined to be such as it is by a like concourse of former causes, may well be called (in respect they were all set and ordered by the eternal cause of all things, God Almighty), the decree of God."[53] And the She-Anchoret explains, "She said, she believed that Jove did order all things by his wisedome, and that his wisedome knew how to dispose to the best, and that Joves will was the onely fixt decree and that his power established all that his will decrees."[54]

Such a power in God, according to Cavendish's fictive persona, is what she understands by the term "predestination." But if all is determined by the fixed decree of God, expressed through the inevitable progression of cause and effect, where is human free will and human responsibility? Both Cavendish and Hobbes attempted to understand this paradox. Hobbes developed a theory of human spontaneity or willingness. For Hobbes, the human striving for self-preservation is basically physiological and necessary because the organism must continue in motion to survive. External objects stimulate the senses by "mechanistic pushes," and the vital, circulatory motion of the heart is either helped or hindered. When the vital motion of the heart is increased, one feels pleasure; when it is diminished, one feels pain. Human reason, differing only in degree from that of other animals, is directly related to our perception of what will cause pleasure and pain, or good and evil. Appetite and fear, stimulated by external causes, alternate during deliberation, which ends in an act of will, either the will to yield (appetite) or the will to abstain (fear). Freedom, or "voluntary action," is merely the ability to act in response to appetite and fear. In other words, one is free whenever there is an absence of impediments to voluntary action; although causally necessitated in one's actions, one acts willingly. Thus, in Hobbes's rendition of mechanistic materialism, the actions of all created beings, from beasts to humans, are determined by the action of matter in motion.[55] Such a doctrine, Hobbes argued, does not excuse or justify human sinfulness: "The nature of sin consisteth in this, that the action done proceed from our will and be against the law. . . . An action may be voluntary and a sin, and nevertheless be necessary."[56]

Cavendish's explanation of human free will and sin was much less straightforward. On the one hand, she denied that human beings can have free will: "For, said she, if Jove had given men Free-Will, he had given the use of one of his attributes to man, as free Power; which, said she, Jove cannot do, for that were to lessen himself for to let any creature have free power to do what he will, for Free-Will is an absolute power, although of the narrowest limits, and to have an absolute power is to be a God."[57] Thus, it would be logically impossible for God to give man the absolute power God alone possesses: freedom is particular to God. Indeed, the She-Anchoret immediately argues that if God had given man free will, it would demonstrate that he was "partial" to one part of his creation over another; either God alone possesses free will, or all being has this power. Moreover, to allow that man could do something apart from God's will "were to make God less than a God, as if his decrees were to be altered by man." On the other hand, it seems that although man does not possess absolute freedom, he still possesses some kind of liberty and is responsible for his own actions and sins. If God required man to do that which a man could not do, "as it were to force him to disobey him or to think Jove suffers man to do evil, when he could prevent it, or to think Jove permits man to provoke his Justice or to damn man, when it is in Joves power to save him, were to think Jove unjust and cruel." Such an arbitrary and

omnipotent God would display a "malignity" in his nature, to damn man for doing what he could not avoid doing, but since God is good, as well as just and
all-knowing, such a thing is inconceivable.[58]

What, then, is this power to act freely in human beings that does not contradict God's absolute majesty? In fact, what Cavendish did in this work and in her
other later works was to incorporate the power of self-movement into the very
essence of matter. Just as disorder was integrated into nature, freedom becomes
an attribute of material being. God is not partial in giving man freedom, since all
of nature possesses freedom, whether from his gift or because of its own internal
composition. Since Cavendish repeatedly described her matter as eternal, and
since she holds that God and nature share corporeality, it may be that in her
theory freedom, God, and matter are consubstantial.

However, in order to locate freedom in the material world, Cavendish has to
explain how matter, and particularly sensate matter, could be independent of
external impact. Hobbes, and Epicurus before him, based their understanding of
freedom and necessity on a physiological understanding of human psychology.
According to Hobbes, in *The Short Tract on First Principles* and *De Cive*, the
latter of which Cavendish had read by 1655, all aspects of being, from the act of
sense (moving bodies impinging on the passive subject) to the instinct for self-
preservation itself, happen necessarily. Neither the soul, the brain, the appetite,
nor the will have any original principle of motion but rather are completely passive and act only in response to an external agent.[59]

Descartes, dealing with the same question in the *Passions of the Soul*, another
text Cavendish had read, had argued for the separation of mind and body and
credited the soul with a kind of volition that we experience "as proceeding directly from our soul and depending on it alone."[60] But since Cavendish denied
the possibility of an incorporeal soul, the Cartesian option was not open to her,
while for someone who wanted to maintain the freedom of the will, the Hobbesian solution was no solution at all.

Cavendish's epistemology of innate matter, which she had first described in
*Philosophical and Physical Opinions*, enabled her to locate complete liberty in the
soul or rational spirits and a more limited kind of freedom in the sensitive spirits.
Rational spirits (or rational matter) largely depend on the mediation of sensitive
spirits (or sensitive matter) to present prints of material objects to the mind, which
then carries commands from the rational spirits, which impress figures on inanimate matter.[61] The process of knowing is a kind of mutual interchange of the
rational and sensitive spirits, whereby the prints or motions received through the
senses are impressed on the mind: according to the She-Anchoret, "the Rationall

Spirits are for the most part Busily imployed in figuring themselves by the sensitive prints, which is the knowledge they take of the works and workings, being more busy and exact, when the sensitive spirits work outward work." But, she explains, the rational spirits are also independent in their actions: "I will not say they move always after the sensitive prints which is to view them, for sometimes they move after their own inventions, for many times the minde views not what the body doth, and many times they move partly after their own invention, and partly after the sensitive prints, but when the sensitive spirits doe retire, or when the rationall spirits perswades them to retire; then the rational spirits move after their own appetites or inventions which are Conceptions, Imaginations, Opinions, Phancies or the like."[62] Free self-movement is the key to all intellectual activities, from conceptions to fancies. The rational spirits can indeed ignore the activity of the senses, which the sensitive spirits must register, "for the sensitive knowledge, which are the sensitive spirits, are bound to parts, but the rational knowledge, which are the rational spirits, are free to all, as being free to it self, the other bound to the dull part of matter."[63]

The sensitive spirits are either directed by the rational spirits or prey to the necessity to pattern the external world observed through the senses. Sometimes, however, the sensitive spirits possess "liberty from those outward labours or imployments: for though they may and are oftimes as active when they work to, or in sleep, yet it is easier being voluntary: for the spirits work more easier, at least more freely, when they are not taskt, then when they are like Apprentices, or Journey-men, and will be many times more active, when they take, or have liberty to play or to follow their own appetites, than when they work, as I said, by constraint, by and for necessity."[64] That is, when the sensitive spirits are free to follow their own wills without impediment, they are free. This kind of freedom is the same kind of freedom that Hobbes postulated for all being. The opposite of willing movement is constraint or necessity, when the sensitive spirits are forced to do what they do not want to do. Sometimes these spirits simply obey, but sometimes they follow the rational spirits "of their own free choice."[65]

Vitalism allowed Cavendish to escape the natural necessity of a mechanistic universe, but she did acknowledge that the close sympathy of the rational spirits and sensitive spirits could discombobulate either. Irregular motion can be a product of the passions of the mind and the violent motion of the rational spirits that "distemper" the sensitive spirits, while the sensitive spirits can "disorder the mind." The most astonishing power of the rational spirits, however, is when their motions cause the sensitive spirits to imprint imaginary sensations and experiences on the senses themselves. The She-Anchoret explains that this can be proved "by those that are affrighted, or have imaginary fear, which see strange and unusual objects, which men call Devils, Hobgoblins, Spirits, and the like: for when men have such imaginary fears they will say they saw strange things, and that they heard strange noises; and smelt strange scents; and they were pinched and beaten black and blue; and that they were carried out of their way, and cast into ditches, or the like; and it is not be doubted but that they did see such sights, hear such sounds, smell such scents, and feel such pains."[66]

There is, therefore, no ontological distinction between real and imaginary being. For Cavendish, every imaginary object is subjectively true when it is generated by the movement of the mind. Those men seeing devils and hobgoblins, or thinking they have been bruised, actually see these apparitions or are black and blue by the power of the rational spirits. Thomas Hobbes dismissed such experiences as "decayed imaginations," which possess no reality: "From this ignorance of how to distinguish Dreams, and other strong Fancies, from Vision and Sense, did arise the greatest part of the Religion of the Gentiles in times past, that worshipped Satyres, Fawnes, Nymphs, and the like; and now adayes the opinion that rude people have of Fayries, Ghosts, and Goblins; and of the power of Witches."[67]

Hobbes may dismiss the superstitions of the common people, but Cavendish does not. In order to substantiate the autonomous power and freedom of rational matter, she has to give credit to even the most fevered imaginings of any person. We know that Cavendish viewed fancy and reason as two complementary operations of the mind: it now appears this union is inherent in the nature of innate matter itself. To Cavendish, fairies, witches, and hybrid beings are substantial and real.

## THE FORCE OF IMAGINATION: TRAVELING TO OTHER WORLDS

The reality of the fantastic is perhaps most obvious in the travel tales that populate *Natures Pictures*, which combine philosophic, literary, and popular notions of other lands and other spheres. In "The Traveling Spirit," a Faustian man seeking wisdom goes to a witch because he wants to see places he could not visit without her assistance. At first, he wants to go to the moon, but the witch tells him that the journey would be obstructed by natural philosophers, who "study Nature so much, and are so diligent and devout in her services, that they despise our great Master the Devil." Then the man wants to go to Heaven but is likewise prevented, this time by the controversies of the Divines, whose arguments so upset travelers they have to turn back from their journeys. Finally, the man asks to go to Hell, but the witch tells him, "I am but a Servant extraordinary, and have no power to go to my masters Kingdome until I dye." He then asks to go to the center of the earth, and after he partakes of opium, his spirit and the witch make the journey.[68]

To some extent this tale is meant as a parody of the usual fantastic tale of travel to areas beyond the borders of the known world. Natural philosophers and divines come in for their share of ridicule. The witch finds the divines particularly odious, which is not surprising given the tale's timing—during the tail end of the European witch craze. What is surprising is that the witch is portrayed as an essentially benign figure, who, regardless of serving the Devil, is happy to help the traveler in any way she can. He is not required to sell his soul to the Devil. This portrayal may be meant as a dig at the witch-hunters and their philosophic supporters, including Cavendish's future correspondents Joseph Glanvil and Henry More, but it may also indicate her sympathy for all kinds of material being.[69] By the seventeenth century, fairies, which had sometimes been associated with witchcraft in earlier times, were increasingly viewed as beneficent.[70] Perhaps for Cavendish, their witchly peers also were rehabilitated.

The fact that the witch takes the traveler to the center of the earth also speaks to this identification of witch and fairy because in traditional folklore that was the home of the fairy queen. Its differentiation from Hell suggests that the earth's center more likely holds fairies than the damned. Indeed, instead of meeting the fairy queen, the traveler encounters an old man, described as a chemist, "who neither stood nor sit, for there was nothing to stand or sit on." He had been hanging at the point that is the exact center of the innermost circle of the earth "ever since the World was made, for he never having a Woman to tempt him to sin, never dyed."[71]

We might describe this figure as Adam's better brother, and indeed he is described as the brother of Adam. While his abode recalls that of Dante's Satan at the very middle of Hell, this eternal being is sinless. Cavendish has reconstrued legends of Heaven and Hell and the accounts of fairyland to allow for a corporeal being that "had the power to call all things on the Earth unto him by degrees, and to dispose of them as he would." In particular, the non-Adam's profession is the most material of the sciences: chemistry. He is the ultimate chemist; unlike his pale alchemical imitators "who made much noyse in talk, and took great pains, and bestowed great costs, to finde the Philosophers Stone, which is to make the Elixar, but could never come to any perfection." Instead, this chemist has created the metals of the earth, the liquids of the earth, and the saltiness of the seas, producing on a grand scale the sulfur, mercury, and salt of the alchemists. Ultimately, he tells his visitor, he will transform the whole earth into glass.[72]

The old man's realm includes Seas of Blood, which are both agitated and calm. They contain the remains of men who have died violently and those who have died peacefully, mingled with the blood of "Beasts, Birds, Fish, and the like" because "the Earth knows no difference" between different kinds of being. Clearly, what they all share is materiality and mortality. Here is an eschatology that at the very least denied the resurrection of the body and asserted the identity of human and animal bodies.[73]

The quest for understanding the nonterrestrial realms continues in the next tale in *Natures Pictures*. In it, a traveler is inspired to learn more about the nature of the air after "he saw something appear in the Air more than usual; which phancy of his caused him to alight from his Horse" and look at the sky. Since he is blinded by the sun, he can see nothing until one of Cavendish's ubiquitous old men gives him a glass with which to observe the celestial globe.[74] This story gave Cavendish the opportunity to explain her optics, which is the same as that described by the She-Anchoret. Vision is caused by species or very small particles streaming into the eye through a small point.[75] This optical theory, as we discussed in Chapter 3, also appeared in the cosmogonic poem in *Poems, and Fancies*, where Life observes species going through a small hole into the concavity of the earth. In this case, the glass allows the traveler to observe the three regions of the celestial globe, which are described as three ascending layers of air and fire. The first region generates the winds and is uninhabited, but the middle and highest regions are populated respectively by fish-men, who live in castles in the air, and salamander-men, who live in noncombustible fiery cities, which are the planets, and are ruled by a king "we call the Sun."[76]

The mixed species that inhabit the celestial regions are among the first of the many hybrids who will populate Cavendish's works, culminating in the man-beasts of the Blazing World. They are all the product of fancy, but fancy with philosophic import. Like the fabulous spirits who are subjectively real in Cavendish's epistemology, fabulous creatures are possible, if not in this world then in some other.

*Natures Pictures* contains one of these stories—"Assaulted and Pursued Chastity"—in which the mental and physical realities of possible worlds are explored. The story of this romance is complex, revolving around a potentially suicidal young lady who, in order to avoid the importunate—and dishonorable—attentions of a married prince, escapes to another land, where she disguises herself as a boy. Happily, the young lady's fate is not our concern, but the beings that capture her in this other land are of considerable interest. The humans she encounters are many colored: the complexions of the lower classes are "deep purple, their hair is white as milk, and like wool . . . their teeth and nails as black as jet, and as shining." The aristocracy and rulers are "of a perfect orange colour, their hair coal black, their teeth and nails as white as milk."[77] The description of the lower classes recalls that of natives in the travel literature of the period, although as Mary Baine Campbell remarks about Cavendish's Blazing World (1666)—which is clearly foreshadowed in "Assaulted and Pursued Chastity"— other worlds function for Cavendish as "a truly alien textual alternative."[78] Nevertheless, even strange new worlds were populated by images drawn from the graphic and literary folklore of the European past. Thus, when Cavendish describes the natives as tall cannibals armed with bows who only wear a kind of diaperlike garment, we should not be surprised by the coincidence with descriptions of the Amerindians in travel literature.[79]

Christopher Columbus wrote of his first voyage, "I have so far found no human monstrosities, as many expected."[80] The expectations of Cavendish's heroine, now named Travelia and dressed as a boy, are not so disappointed. In addition to the humans she sees, she encounters strange beasts, which are hybrids of many animals. So, for example, she is carried from place to place on the back of "a creature half fish, half flesh, for it was in shape like a Calf, but a tail like a Fish, a horn like a Unicorn."[81] These mer-beasts, who unite unicorns, monstrous calves, and mermen, signify that the heroine has entered a world of wonders, but it is still a world made up of the prodigies that thronged the popular culture of the age.[82]

The strange beings of other lands and other worlds provoked much philosophic and theological discussion in the sixteenth and seventeenth centuries. The central focus of this discourse was the examination of whether natives, either in America or in Africa or on the Moon, possessed immortal souls, or even whether they were human. It is not surprising that when Travelia, disguised as a boy, tries to dissuade the king and people of this imaginary world from sacrificing her, she speaks of the soul.

This speech includes an almost word-for-word restatement of the She-Anchoret's doctrine of the nature of the soul (rational matter) and its relationship to sensitive and inanimate matter. But it is also a sermon forecasting salvation or damnation. Travelia is a preaching woman (although described as male) and as such closer to the radical women of the Commonwealth sects than to the virgin saints of medieval hagiography.[83] She informs the natives that the gods both punish and reward men and "their love not only saves man, but prefers man to a glorious happiness." And "though it [the soul] is not a god from all eternity, yet it is a kind of deity to all eternity, for it shall never dye." She asks her persecutors if they would have the temerity to wrench the soul away from the sensitive matter "before it be the Gods pleasure to dissolve that body, and so remove the Soul to a new Mansion." Thus, Cavendish implicitly endorsed the doctrine of the immortality of the soul, but she was ambivalent about whether every man—white, purple, or black—may possess it. Travelia tells her listeners, "And although it is not every Creature that hath that Soul, but onely Man, for Beasts have none, nor every Man, for most men are Beasts, onely the Sensitive spirits and the Shape may be, but not the Soul; yet none know when the Soul is out or in, but the Gods, and not onely other Bodies may not know it, but the same Body be ignorant of thereof."[84]

The possession of the soul is therefore not species-determined: most men share the bestiality of animals and lack the humanity the soul endows. Consequently, in a kind of Pascalian bargain, "you must seek all the wayes to preserve one another" because one does not know who has soul and who does not. The alternative for those who kill is "perpetually dying and killing with all manner of torments" as they pass from one kind of body into another in a reincarnation of bodily misery, which afflicts the mind as well as the body: they die "thus burning, hanging, drowning, smothering, pressing, freezing, rotting, and thousands of these kinds, nay more than can be reckoned, may suffer: thus several Bodies, though one Mind, may be troubled in every Shape." Those others, however, who please the gods, "live easy in every Shape, and dye quietly and peaceably; or when the gods do change their Shapes or Mansions, 'tis for the better, either for ease or newness."[85] Thus, whether body or soul survives, the fate of each individual is clearly material.

What the ontological and theological status of the soul might be is examined in another part of *Natures Pictures*, when a man seeks to know about the fate of the soul:

> There was a Man which much desired to know,
> When he was dead, whether his Soul should go;
> Whether to Heaven high, or down to Hell,
> Or the Elizium Fields, where Lovers dwell;
> Or whether in the Air to flee about,
> Or whether it like to a Light goes out.[86]

This litany of the possible fortunes of the soul after death recalls Christian, pagan, and Epicurean outcomes. The significance of Heaven and Hell and the Elysian Fields is clear in this passage; the reference to the fleeing soul refers back to the atomism Cavendish rejected in "A Condemning Treatise": "I cannot think that the substance of infinite matter is only a body of dust, such as small atoms . . . nor [there is] no change and variety, but as they move, as onely fleeing about as dust and ashes."[87] The allusion to a light going out may recall the Stoic doctrine of the soul being a part of the divine fire.[88]

It seems there is no definitive answer to the question of what happens to the soul, but Cavendish nevertheless keeps asking. She cannot construct a natural religion without considering it, and in "The She-Anchoret" Cavendish suggests that "in nature there was a Hell and a Heaven, a God, and a Devill, good Angells and bad, salvation and damnation." But this is only an earthly fate, where "pain and trouble is a Hell, the one to torment the body, the other the mind," while "health and pleasure is a Heaven, which gives the body rest, and the minde Tranquillity."[89] This solution may have pleased Epicurus, but it would not have satisfied the believers of any Christian religion. And to compound the problem, Cavendish introduced the idea in this text that the soul itself may belong to animals as well as humans; the boundary between man and beast wavers: humans may be soulless and beasts soulful. The moral philosophers had asked the She-Anchoret "if she thought beasts had a rationall soul." She answered

> that if there could be not sense without some reason, nor reason without the senses, beasts were as rationall as men, unless, said she, reason be a particular gift, either from nature, or the God of Nature to man, and not to other creatures, if so, said she, Nature or the God of Nature would prove partiall or finite; as for Nature in her selfe she seemes unconfined, and for the God of Nature, he can have no byas, he ruling every thing by the straight line of Justice; and what Justice, nay, injustice would it not be for mankind to be supreme over all other animall kinde? Some animall kinde over any other kinde?[90]

God's partiality is again the issue as it was in the question of human free will. In this case, partiality implies finitude, but God's justice necessitates the equality of all created being, including animals and man. Likewise, the dominion of man over beast, or beast over beast, would be a sign of God's injustice, which by implication is impossible.

On the ontological level, it is the vitality of innate infinite matter, created by an infinite just and good God, that results in a doctrine of the permeability of human and animal kinds. In a sense, matter itself is hermaphroditical: a combination of forms of rational, spiritual, and inanimate principles. Although the principles of nature are inseparable, its concretions can be more or less rational or spiritual or inanimate; thus, men can be beasts and animals can be men.

In *Natures Pictures*, Cavendish attempted to answer the most difficult metaphysical, moral, and religious questions inherent in her vitalistic materialism. She manipulated gendered metaphors of nature and God to discuss the existence of disorder and evil in nature. She considered the possibility of freedom of will and determinism in a universe composed only of matter in motion. She speculated on the constitution of the soul itself, its ultimate fate, and whether it was particular to human beings. She developed her epistemology and psychology to allow for the reality of subjective conceptions: anything that can be conceived of can exist. Many of the stories contained in *Natures Pictures* testify to an imagination inspired by travel literature, where travelers, both male and female, encounter strange phenomena and hybrid beings. The substance of Cavendish's tales, like the matter of nature, subsumes a reality where anything is possible.

After completing *Natures Pictures* in 1656, Cavendish did not publish again until the 1660s. In the intervening time, she became much more serious about studying the philosophies of her contemporaries, in part because she wanted to be treated seriously by them. Increasingly, she came to view her natural philosophy as not just possible but also probable and fantasy as an adjunct to but not a substitute for reason. As her material philosophy matured, she realized that she had to confront the ideas of others while also proclaiming her own originality. Her doctrines of innate matter and the soul, God and nature, freedom and necessity, and particularly the hybrid mingling of species put her at odds with Hobbes and Descartes, Van Helmont and Henry More. Her sympathy with nature and animals, and a broadening conception of the presence of rational matter in all being, caused her to challenge these philosophers and the Royal Society. Her observations of nature were complemented by her observations of society and politics. In these later works, Cavendish was willing to engage others and eager to provoke their attention in turn. Cavendish created a place for herself and her doctrine in the subjective reality of her mind, and she wanted external as well as internal validation of her work.

# chapter five

# The Politics of Matter

*B*etween the publication of *Philosophical and Physical Opinions* in 1655 and its revision in 1663, another world opened for Margaret Cavendish. In 1660, Charles II was restored to the English throne, and his loyal followers returned to England. Newcastle set to work to restore his ruined estates, and the king rewarded his loyalty by making him a duke, although without admitting his old governor to his personal circle. For the rest of their lives, Newcastle and Cavendish lived largely in retirement at Welbeck Abbey, except for an occasional trip to London. This retreat could not have been distasteful for Cavendish, as she achieved the solitude her projected other self had gained in "The She-Anchoret." The poet Richard Flecknoe described her during this period:

> What place is this? looks like some Sacred Cell
> Where holy Hermits anciently did dwell,
> And never ceas'd Importunating Heav'n
> Till some great Blessing unto Earth was giv'n;
> Is this a Ladys Closet? 'tcannot be,
> For nothing here of vanity we see;
> Nothing of Curiosity nor Pride,
> As all your Ladies Closets have beside;
> No mirrour here in all the Room you find;
> Unless it be the mirrour of the Mind,
> Nor Pencil here is found, nor Paint agen
> But only of her Ink and of her Pen.
> Which renders her an Hundred times more fair
> Than they with all their Paints and Pensils are:
> Here she is Rapt, here in Extasy
> VVith studying high and deep *Philosophy*.[1]

The return to England allowed Cavendish the time she needed to study the writings of other modern natural philosophers, several of whom she critiqued in *Philosophical Letters* (1664).[2] She was prompted to reconsider her materialism after critics charged that *Philosophical and Physical Opinions* was neither philosophic nor original. Her exasperation with her critics is clear from her citing Aesop's tale of the father and son criticized by townspeople for how they rode their donkey— singly, together, alternately, and not at all. Finding that they could never please anyone, the father resolved to drown his animal, but Cavendish was "not so passionate to burn my writings for the various humours of mankind."[3] Nevertheless, she resolved to revise, correct, and clarify her *Opinions* in order to build "not only a Larger, but more Exact and Perfect Fabrick, wherein every Several Chapter, like Several Rooms, have as Much and Clear Light as I can give them."[4] In 1663, she published a much expanded edition of *Philosophical and Physical Opinions*.

Most significant, in the later edition of *Philosophical and Physical Opinions* (henceforth referred to as the 1663 *Opinions*), Cavendish developed an expanded theory of matter, which emphasized its unity but nevertheless maintained the hierarchy of different kinds of matter, now so entirely integrated that her matter theory can be described as holistic. Her aim was to explain how matter—or people—can be arranged in order to avoid disorder and achieve a harmonious state of nature or society.

Most of the development and revisions in this work concern matter, motion, and the nature of man—physically and psychologically.[5] The changes in the new edition included a more mature and nuanced description of her own materialism, emerging not only from her more sophisticated knowledge of other natural philosophies but also from a deepening concern with her philosophy's political implications. Many of the speeches in *Orations of Divers Sorts* focus on political ideas, including the position of women. Her 1662 edition of *Plays* also contains many political themes. Cavendish's political philosophy eludes easy categorization; although she is often characterized as a monarchist, she has also been linked to radical republicanism.[6] One way to understand her political theory is to relate it to her natural philosophy. Scholars connect Thomas Hobbes's natural and political philosophies; Cavendish's very different ideas about nature and politics also are interdependent.[7] She rejected Hobbesian determinism and absolutism, but she understood that matter in motion included the actions of human beings within the state.

Hobbes, her husband's old friend, thought that order in nature was the result of the impact and entanglement of matter in motion; order in the state could occur only when colliding individuals gave their collective power to an absolute ruler. Cavendish argued instead that just as every part of material nature—rational, sensitive, and inanimate—cohered together and functioned as a whole, so every member of a well-ordered polity naturally unified to create a strong state, with each constituent functioning to perform its own duties. Hobbes emphasized the artificial beginnings of the state; Cavendish argued that humans, since they were composed of rational and sensitive matter, always lived in a political state. Although the state was sometimes disordered by rebellion and civil war, the natural state of humanity and nature was harmonious—by its nature and not by artificial construct.

Cavendish's revised vitalistic materialism, which now included a kind of holistic concept of matter, emphasized harmony and unity of matter without completely abandoning her earlier depiction of the struggle and potential disorder inherent in motion. In the 1650s, when Cavendish speculated on political themes, she often emphasized the turmoil that could undermine both the state and nature. By the 1660s, Cavendish's materialism reflected England's more stable political situation and her belief in traditional hierarchies and monarchical rule. But the recent experience of the civil war, and her husband's dismissal from the center of power, had disabused Cavendish of any complacency about politics. She felt that just as the different but united parts of matter should be given responsibility reflecting their capacities and place in the scheme of things, so absolute rulers should rely on their traditional supporters, the nobility, and on the ancient traditions of law and government. Nevertheless, her sense of alienation is reflected in some extraordinary speculations about the role of women in the state in her nonscientific works, which are potentially as radical as Hobbesian absolutism. But she dissociated herself from Hobbes in *Philosophical Letters*, where she presented her belief in the possibility of free action in the state and nature and disputed the determinism of Hobbes's system.

## THE HOLISTIC MATERIALISM

In 1663, citing lack of clarity in the 1655 edition, Cavendish decided to be more precise in defining the metaphysical terms of her natural philosophy, including one of its most opaque aspects: the nature of individuated and differentiated parts of matter. While this was less of a problem in *Poems, and Fancies*, which embraced the atomistic notion of indivisible smallest parts ordered by external principles, it became more of an issue when Cavendish adopted the theory of innate matter. Cavendish declares that matter possesses quantity, shape, and motion and that it is rational, sensitive, and inanimate, but what is the relationship of any piece of matter to any other piece of matter? If all matter is self-moving, how is the order and unity of nature achieved without introducing a God who determines how matter will move, thereby sacrificing liberty for providence? In 1655, Cavendish believed she had solved this problem simply by asserting that matter is inherently self-moving and self-determining, but by 1663 she felt she needed to elaborate her explanation.

The subtle but pervasive role of political considerations in the development of Cavendish's ideas is most clear in her revised explanation for repudiating atomism. In "A Condemning Treatise of Atomes" (1655), she based her reasons for rejecting mechanistic atomism largely on the physical disorder such a fortuitous system would possess: "But as they move, onely by fleeing about as dust and ashes, they are blown about with winde, which me thinks should make such uncertainties, such disproportioned figures, and confused creations, as there would be an infinite and eternal disorder."[8] In 1663, this criticism was repeated, but Cavendish added the following, "Unless every single Atome were Animated matter, having Animated Motion, which is Sense and Reason, Life and Knowledge, to Move and Create other Figures, by Joyning and Uniting their Small Bodies by Consent, and Dissolving by Consent and Agreement, which is not Probable."[9]

Up until the "which is not Probable," it seems we are on familiar ground. Innate matter, the core of her earlier natural philosophy, possessed just the characteristics Cavendish here rejected. It combined sense (sensitive matter) and reason (rational matter), and life (sensitive matter) and knowledge (rational matter); it was a self-moving and self-determining principle of being. But the liberty that was essential to innate matter now becomes a detriment:

> For if Every and Each Atome were of a Living Substance, and had Equal Power, Life and Knowledge, and Consequently a Free-will and Liberty, and so Each and Every one were as Absolute as any other, they would hardly agree in one Government, and as unlikely as Several Kings would agree in one Kingdom, or rather as Men, if every one should have an Equal Power would make a good Government; and if it should Rest upon Consent and Agreement, like Human Governments, there would be as many Alterations and Confusions of Worlds, as in Human states and Governments by Disagreement, for there must necessarily be as much Liberty and Power in every Atome to Disagree as to Agree.[10]

The analogy between atoms and men having equal power to create governments or worlds "perswades me to Wave the Opinion of Atoms." But whose opinion of atoms is Cavendish waiving? It appears that Cavendish is rejecting not only Epicurean atomism but also her own vitalistic materialism. Her dislike of political autonomy has trumped her philosophy. It simply would not do to endorse a physical system that could be used to justify the most radical republicanism of the Commonwealth period, where each individual is autonomous before uniting to form the state.[11]

But it may be that Cavendish understood that her works could be read in such a way and she consequently disavowed the autonomy and self-rule suggested in the vitalistic materialism of the 1655 *Opinions*. Thus, in the second edition of *Philosophical and Physical Opinions,* Cavendish expressly rejected a matter theory of autonomous, individuated parts of matter and instead developed the idea of matter as one extended and infinite substance divided into parts only by its own motion. These parts do not exist independently of the whole but are simply aspects of its being. In a somewhat Parmenidean picture of the universe, Cavendish explained, "No perfect Division can be made in Infinite matter, as to Divide one part from the rest, for though Parts may be made as Folds, or into Figures, and these to Remove from place to place, and Parts to or from Parts, yet they cannot be divided from the Whole, as to become each a Single part of it Self; but they remain still Parts, pertaining to the Only matter, for though there may be Infinite parts of infinite matter, yet there are not Infinite Single parts, but Infinite Inherent parts."[12] Matter and its divisions are one: its parts are a part of its body. Thus, fusion and unity here become the guiding principle of Cavendish's metaphysics

Cavendish's holistic materialism is elaborated in several places in this text and will be the core of all her later philosophic writings.[13] Through it, Cavendish integrated her former hierarchy of matter into the structure of the universe, without sacrificing unity for diversity. Matter continues to be characterized hierarchically as rational, sensitive, and inanimate, but "these several degrees alter not the nature and entity of Only matter, but if part of Only matter were not Animate, there would be no Motion, and if there were no Unanimate matter, there would be no Gross Substance, and if there were no Degrees in Only matter, there would be no Change or Entercourse in Only matter, or in the Nature of Only Matter."[14]

In this version of her material philosophy, innate matter is renamed animate matter, still divided into two parts, rational and sensitive. The dull part of matter is still "unanimate" or "inanimate matter." Rational matter is the director, and sensitive matter the worker shaping inanimate matter—and the moving life force that mediates between the mind or rational matter and the inanimate body of nature—but all aspects of matter are one, integrated and infinite. They are "a prime triumvirate of the prime degrees of the Only and Infinite matter, living, mixing, and moving together, as the Body, Life, and Soul."[15] Cavendish argued that since matter is infinite in all its parts it is therefore "intermixed" and a unity.

Cavendish repudiated mixing natural philosophy with theology (her fideism is very explicit in the 1663 *Opinions,* although the religious prefaces from the first edition are not reprinted), but here she articulated a matter theory recalling the Christian trinity, where three elements are actually one. A year later, in *Philosophical Letters* (1664), she expressly applied this analogy to her philosophy's three qualities of matter: matter/soul (rational matter), motion/life (sensitive matter) and figure/ body (inanimate matter). She invented another self (Lady N. M.), who in a letter to a third self (Madam) inquired of her: "I desired, if she could not make me understand the mystery she would but inform me, how three made one in Divinity, Nature, and Man. She said, That was easie to do; for in Divinity there are three Persons in one Essence, as God the Father, the Son, and the holy Ghost, whose Essence being individable, they make but one God; And as for Philosophy, there is but Matter, Motion, and Figure, which being individable, make but one Nature; And as for Man, there is Soul, Life, and Body, all three joyned in one Man."[16]

Fundamentally, the triune theory of matter allowed Cavendish to picture a universe in which authority and freedom reside in a ruler who operates through an assistant in order to control a gross populace, without sacrificing the unity or order of the state. Rational matter is "pure and free," and sensitive matter is "the architectonical part of Infinite matter, to fit and form the Unanimate part of Matter into Figures."[17] Harmony prevails.

In the first edition of *Opinions,* Cavendish emphasized the degree of discord in nature, with each motion in constant strife with other motions for power, and matter disputing with infinity for control of itself and sometimes becoming a tyrant to itself. This passage is reproduced in part 6 of the 1663 *Opinions,* with only the terminology describing matter changed. But earlier in the text, when Cavendish describes her theory of holistic matter, she softens the description of the discord in nature. While "Motion, the Creator of Figures, doth make Warr," animate matter normally constructs a peaceful universe: "But Motion, the Active part of Infinite matter, would cause Confusion, being in its nature restless; but the Only matter, being Intire in its own nature, also Infinite and Eternal, is the cause that Confusions cannot be. . . . In truth, the Unity of the nature of Only and Infinite matter, maketh Concord out of Discord, so as the Cause ordereth the Effects, for the Effects cannot alter the Causes, for though the Effects make Disturbance, the Causes make Peace."[18]

Cavendish had always posited a degree of harmony between rational and sensitive matter, but now the potentially discordant motion possessed by both parts of animate matter can be controlled through the unified substance of matter. The potential for disorder (and civil war) always remains; the different parts of matter can disturb each other. But normally, in different beings, "there is a strong Sympathetical Agreement, and Natural Unity between the Rational and Sensitive Matter and motions in one and the same Figure and Creature, not only for the Figures or Creatures Consistence, Maintenance, Use, Ease, Pleasure, and Delight, but for the Guard, Safety, and Defence." This concord depends on the fact that "the Rational advises, the Sensitive acts."[19]

Unity and hierarchy are not antithetical in Cavendish's natural or political philosophy. She is neither an absolutist nor a republican. Rather, her theory of nature and the state reflects traditional ideas about the functional importance and public responsibility of each order. A truly strong state, or truly strong matter, needs to integrate its different constituents: "It is with the Animate matter and motions, as it is with Governours, and Citizens, or Commons, they know not their own Power and Strength, until such time, as they make a Trial, or are Forced to it, for every Particular part knows not the Strength of the Whole, untill the Several Parts joyn all into One."[20]

Sometimes "Uproar, or Tumult" characterizes the state and matter, but if every part of the whole recognizes its particular function, "Strength and Power" result. In fact, in Cavendish's other works of the early and mid-1660s, including *Orations*, *Blazing World*, and her 1667 *Life* of her husband, she suggested that the reason a monarch may get into trouble is because he allows courtiers and favorites to usurp the traditional roles of wise statesmen and councillors and allows them to oppress the people.[21] Since peace and order are the goal of everybody, the parts who rule and the parts who obey must be integrated.

## POLITICS AND FANCY

Cavendish was by no means a systematic political thinker. The multiplicity of genres in her writings reflected experimentation with different forms of political organization. It also allowed her a forum to express potentially subversive political opinions. She examined the rights and power of kings and the corresponding obligations of male and female citizens. She sought the source of disorder in the state and proposed solutions to avoid civil war. In her works of the 1650s, she considered the functions of the different members of the state and imagined commonwealths formed and maintained through social contract, an idea rooted in medieval constitutionalism but finding a new life in the religious and political upheavals of the sixteenth and seventeenth centuries. By the early 1660s, perhaps reflecting the effort by Charles II to regain royal prerogatives, she was more inclined to emphasize the power of the ruler. But in all her works, she insisted on the rights and liberties of subjects/citizens, the importance of hierarchy and order in the state, and obligation of the king to rule justly.

In *Poems, and Fancies*, Cavendish imagined a cosmogony where a female ruler, Nature, aided by her four lieutenants—Matter, Motion, Figure, and Life—created the cosmos. Although motion sometimes caused disorder, Cavendish depicts a smoothly running commonwealth, ordered and commanded by a ruler who is powerful but not absolute. Death, vacuum, and fortune all defy her power, but most of nature is well ordered. In another fable in *Poems, and Fancies*, at the very end of the text, Cavendish shows animate being as a state ruled by a powerful sovereign, who nevertheless consults with his people before making law. In "The Animall Parliament," she analogized the human body to a state, reversing the usual political metaphor of the body politic in which every part of the state is represented as parts of the human body: "The Soul called a Parliament in his Animal Kingdom, which Parliament consisteth of three parts, the Soul, the Body, and the Thoughts, which are Will, Imaginations, and Passions. The Soul is the King, the Nobility are the Spirits, the Commonality are the Humours and Appetites. The Head is the upper House of Parliament, where at the upper end of the said House sits the Soul King, in a Kernal of the Braine, like to a Chaire of State by himselfe alone, and his Nobility around him."[22]

In this body, there is a lower house of Parliament also, composed of the knights, gentry, and burgesses, and all of the senses and passions are represented. It may seem odd to find a Parliament in a work written by a royalist woman at a time she was seeking funds from a Parliament that had just beheaded its king. But "The Animall Parliament" is a fable of how the state should work, not how it was working. Its king, recognizing the potential of "riotous disorders" in the state, determines to create a kingdom marked by "Justice, Prudence, Fortitude, and Temperance" and "to make and enact strict Lawes to a good Life, in which I make no question, but every one which are in my Parliament will be willing to consent, and be industrious thereunto."[23]

Cavendish's belief in strong monarchy did not preclude a role for the citizens of the state. There is a difference between a good king and a tyrant: "For I desire to be the King of Affection, ruling them with Clemency, rather then to be only King of Power, ruling them with Tyranny, binding my Subjects to slavery." And so in the fable, the soul of the body listens to the concerns of his citizens and rectifies the troubles in the state of the body, and the citizens gladly consent to the taxes he requires. This is an idealized picture of how both nature and government can be well ordered—not the case in either an absolute monarchy or a tyranny.

Cavendish repeatedly returns to the issue of monarchial power. In *The Worlds Olio*, published in 1655 but the first work she wrote, Cavendish depicted a Commonwealth "composed of Nobility, Gentry, Burgesses, and Pezants . . . governed by one Head or Governour." The first article of their social contract is "that this Royal Ruler to swear to the People to be Carefull and Loving, as well as the People swear Duty and Fidelity." The mutual obligations that follow are essentially Cavendish's indictment of the failed government of Charles I. Rulers, she suggests, should never have favorites or sell offices but rather should "reward the Meritorious, and grace the Virtuous." Clerics in this state should not "dispute" but preach, and the different ranks should be distinguished by appearance and ceremony. All citizens should pay their taxes, and the monarch should respect their property rights and control commerce. In society and the state, all due order and reverence should be upheld between superiors and their dependents: "No Children shall speak before their Parents, no Servants before their Masters, no Scholars before their Tutors, no Subject before the Prince." Interestingly, within the matrimonial commonwealth, the claims of the dependents on their superiors is emphasized: "All Husbands shall use their Wives with Respect, unless they dishonour themselves with the neglect Thereof."[24]

In the 1655 *Opinions*, the harmony and good order of material nature were also depicted in a political metaphor. One passage—which at its beginning makes Cavendish sound like a utopian republican—ultimately emphasizes the hierarchy and coherence of matter and the state: "There are millions of several motions which agree to the making of every figure, and millions of several motions are knit together; for the general motion of that are figure, as if every figure had a Common-Weale of several Motions working to the subsistence of the figure, and several sorts of motions, like several sorts of Trades hold up each other; some as Magistrates, and rulers; others as Train-bands [citizen soldiery], or souldiers; . . . some as Merchants that traffick; . . . some that labour and work."[25] Working together, like the middling sort in a well-run state, motions create a harmoniously shaped figure. But ultimately, in this text, the hope for order in nature is defeated by "matter running perpetually towards absolute power."[26] Nature often faces discord, just as the good monarchy Cavendish had hoped for was destroyed by civil war. Neither party in the social contract has upheld its end of the bargain. The sensitive matter sometimes rebels against the rational matter, and the rational matter sometimes agitates the sensitive matter.

Many of the political themes Cavendish discussed in *The Worlds Olio* and alluded to in her natural philosophy reappear explicitly in *Orations* (1662), which contains persuasive speeches mostly concerning political topics. Cavendish frames her rhetorical exercise as an exercise in imagination for both her reader and herself; "first imagining my Self and You to be in a Metropolitian City," she invites the reader to the marketplace, the Court and Council table, the law courts, the grave-yards and the weddings, and "Private Conventicles" of women.[27] As in many of her dialogic works, in these imagined places Cavendish presents opposing points of view that are sometimes resolved in compromise. It is, therefore, difficult to say which argument she favored, but one conclusion is that Cavendish was very famil-iar with the discourse of political theory characterizing early modern Europe, even while she claimed women had no role to play in politics. Perhaps when she re-turned to England with the Restoration, Cavendish perused political texts as well as philosophic treatises. *Orations* echoes the discussions of absolute monarchy, re-publicanism, and the ancient constitution of England. The entire work is in the "mirror for princes" tradition, in which political thinkers—from Sir John Fortescue to Thomas More to her own husband—advised monarchs on how to rule justly and successfully. Speaking through the voice of a Privy Counsellour, Cavendish writes, "I think it is my Duty . . . to give your Majesty Advice, lest Sudden Dangers may Surprize you, or at least great Disorders may give you great Troubles."[28]

The most important function of government—whether monarchical, aristo-cratic, or democratic—is to avoid "confusion" and the destruction of the state. In "A Souldiers Oration concerning the Form of Government," the soldier argues, "But one of these Governments must Settle in, otherwise all the Kingdome will be in a Confusion; for if there be no Order and Method, there will be no Rule nor Government, since Every one will do what he list."[29] As in an atomistic uni-verse, without an ordering principle, disorder—or civil war—reigns.

The immediate context of Cavendish's political writings is the English Civil War. She was fully cognizant, as she had been in *The Worlds Olio*, that the rebel-lion occurred, at least in part, because of abuses of power under Charles I. Thus, in an oration entitled "Disorders, Rebellion, and Change of Government," the speaker argues that civil war was caused by "Flattery, Vanity, and Prodigality" and "the Selling of all Offices and Places of Judicature." Corruption caused "great Taxes laid upon the People and Kingdome," and the result was a republic in which there was constant contention and ultimately civil war. The speaker pro-claims, "O Powerfull Voice of a Headless Monster!"—to which he suggests the only ultimate recourse is the rule of a "Native King . . . that we may See our own Errors and Reform our Faults, and herafter Live Happily under the Government of a Good and Wise King, which I Pray the Gods Send to you."[30]

Many of Cavendish's orations conclude that the best government is a monar-chy reflecting the ancient composition of the state. In a discussion of liberty of conscience, one speaker argues that "there ought to be a strict Law, that no Governour or Magistrate shall in any kind Infringe our Just Rights, our Civil or Common Laws, nor our Ancient Customs; for if the One Law should be made and not the Other, the People would be Slaves, and the Governours their Tyrants."[31] In a speech in which a king is trying to mollify his discontented subjects, the monarch contends, "To make War on the Protector of your Liberties, and Father of your Country, is Unnatural," and in yet another speech discussing the same subject, the orator proclaims, "O Foolish People, that will quit your Present Happiness for a Voluntary Slavery! and as for a Monarchical Government, which you seem to be Weary of, it is the most Ancient and Divinist, as being an Imitation of God and his Angels in Heaven, wherein are Degrees, as Higher and Lower from, and to his Throne."[32]

Monarchies are ancient and closest to the divine. The good ruler protects the liberties or rights of his subjects; his rule is not arbitrary, and his people are divided into ranks or degrees. The importance of hierarchy informs Cavendish's politics as it did her materialism. In another oration, the king is equated with the brain, which directs the nobles, who are the head, who in turn govern the body or the common people.[33] Here again we meet rational, sensitive, and inanimate matter transposed into a political metaphor. This image differs from the traditional body politic analogy by equating the nobles with the head rather than the arms of the body; these nobles "Guide, Direct, Rule, and Govern" in concert with a king who is the "Chief Governour." "Wherefore," the speaker concludes, "a King or Chief Ruler, Joyn'd to a Grand Counsel, is the Best Government of all."[34] A "mixt" government, integrated in the same way as a harmonious body, precludes civil war and creates the best-ordered state.

In these orations, Cavendish comes close to mimicking the arguments of Sir Edward Coke in the early seventeenth century about the ancient constitution of England. Defending the rights of Parliament against the attempted absolutism of James I's government, Coke claimed that the king must govern in accordance with the common law and respect the property and traditional rights of his subjects. Cavendish clearly was familiar with this debate and other controversies about political authority in early modern Europe. Some of her speakers endorse the theory of the divine right of kings, while others argue for the patriarchal origin and justification of monarchy.[35] A Privy-Councellour in one oration even paraphrases Machiavelli's views on whether a prince should govern through love or fear. He states, "But when they [the people] Fear their Soveraign, they are Obedient; for it is impossible to Work upon their Good Nature, as to make them Obey through Love and Good Will, because they have no Good Natures to work on." Consequently, "there is none other way but Force, to make them Loyal and to keep them to their Alligiance."[36] Monarchy, ruling through compulsion, becomes the best way in this case to avoid the terrors of civil war.

Just as Cavendish was willing to contemplate the most absolute forms of government, she also presented the arguments for republicanism, democracy, and even forms of communism, echoing the theories of the Diggers and other radicals during the civil war. A lawyer argues, in defending a thief, that "Nature, who made all things in Common, She made not some men to be Rich, and other men Poor . . . for when she made the World and the Creatures in it, She did not divide the Earth, nor the rest of the Elements, but gave the use generally amongst them all."[37] This creator Nature, who sounds much like the Nature described in Cavendish's early natural philosophy, "is the Empress of Mankind, Her Government is the Ancientest, Noblest, Generousest, Heroicest, and Royalest." The laws of Nature, therefore, guarantee all men "Natural Liberties and Inheritances, which is to be Equal Possessors of the World," and when some "Moral Philosophers, or Commonwealth makers" instituted laws and private property, they "were Rebels against Nature, Imprisoning Nature within the Jail of Restraint, Keeping her to the spare Diet of Temperance, Binding her with Laws, and Inslaving her with Propriety."[38]

The actions of lawmakers parallel and anticipate the actions of experimental philosophers who also try to control and enslave Nature. The speaker who argues against the view that Nature's power has been usurped by the founders of government disputes the image that Nature is lawful and benevolent: "There is no Law in Nature, for Nature is Lawless, and hath made all her Creatures so, as to be Wild and Ravenous, to be Unsatiable and Injurious, to be Unjust, Cruel, Destructive, and so Disorderous, that, if it were not for Civil Government, Ordained from an Higher Power, as from the Creator of Nature her self, all her Works would be in a Confusion, and so their own Destruction."[39] In a state of nature, where there is no law, it sounds like there is a war of all against all until laws are imposed by the state. But in this oration, Cavendish steps back from a completely Hobbesian solution to natural anarchy by introducing the divine into the calculation: the gods give man a "Supreme Power . . . to Rule and Govern Nature."[40] Government is not a human construct but a divine artifact.

Cavendish was certainly willing to contemplate radical political philosophies, if not ultimately to endorse them. In *Orations*, she argues the pros and cons of liberty of conscience and concludes with the pragmatic and *politique* solution that as long as religious beliefs don't interfere or disrupt the state, "let them have Liberty of Conscience."[41] The most important aim for any political or indeed social entity is to ensure peace and order. This ultimate aim leads another orator to suggest that women be restrained from public and even private meetings because of their potentially disruptive character: "Femal Societies . . . ought to be Dissolved, allowing no Publick Meetings to that Sex, no not Child-bed Gossipings, for Women Corrupt and Spoil each other. . . . Liberty makes all Women Wild and Wanton."[42] Presumably Cavendish does not agree with this sentiment, and it is immediately followed by an argument for female liberty of association, which perhaps would make a modern feminist cringe. "Women were made by Nature for Men, to be Loved, Accompanied, Assisted, and Protected," another male orator proclaims, and therefore should be "Beloved, Admired, Desir'd, Ador'd, and Worshipp'd, Sued and Praised to by our Sex."[43] This sentiment equally objectifies women, but at least in a complimentary way.

Orations and the other nonscientific texts written in the early 1660s present a complex view of women. The debate about the rationality and nature of women follows the discussion of their liberty and position vis-à-vis men. For the only time in Orations, women speak. We have already heard some of their arguments. The first speaker argues that women should "make a Frequentation, Association, and Combination amongst our Sex, that we may Unite in Prudent Counsels, to make our Selves as Free, Happy, and Famous as Men," while another agrees that men "keep us in the Hell of Subjection" but despairs, "I cannot Perceive any Redemption or Getting out; we may Complain, and Bewail our Condition, yet that will not Free us. . . . Our Power is so Inconsiderable, as Men Laugh at our Weakness."[44]

Women are powerless, but another orator claims that "we have more Reason to Murmur against Nature than against Men, who hath made Men more Ingenious, Witty, and Wise than Women . . . for Women are Witless, and Strengthless, and Unprofitable Creatures, did they not Bear Children."[45] Unhappily, from the modern perspective, most of the other women in the discussion agree with this sentiment about the natural inferiority of women, although they try to recast it as advantageous in allowing women to live more protected and peaceful lives. What all this means about Cavendish's attitude toward female participation in the state is therefore left up to the reader.

Perhaps a comment in Sociable Letters best expresses Cavendish's evaluation of the position of women in the political state. She writes,

And as for the matter of Governments, we Women understand them not, yet if we did, we are excluded from intermeddling therewith, and almost from being subject thereto; we are not tied, nor bound to State or Crown; we are free, not Sworn to Allegiance, nor do we take the Oath of Supremacy; we are not made

Citizens of the Commonwealth, we hold no Offices, nor bear we any Authority therein; we are accounted neither Useful in Peace, nor Serviceable in War; and if we be not Citizens in the Commonwealth, I know no reason we should be Subjects to the Commonwealth: And the truth is, we are no Subjects, unless it be to our Husbands.[46]

Cavendish here argues that, disenfranchised from the state, women are not subject to it. They apparently live in a state of nature outside the bounds of law. But bountiful Nature intervenes, giving women a covert power over their husbands: "They seem to govern the world, but we really govern the world, in that we govern men, for what man is he, that is not govern'd by a woman more or less? . . . They are Led, Guided and Rul'd by the Feminine Sex."[47] Female power is covert, and rooted in manipulation, but power nevertheless.

But in fiction, women can have overt as well as covert power. In her plays, Cavendish was more willing to contemplate radical political roles for women, where they routinely transgress gender norms because men have failed in their duties. In "The Unnatural Tragedy," sociable virgins propose that women take over the rule of states:

> 3Virgin: Certainly, if we had that breeding, and did govern, we should govern
> the world better than it is.
> 4Virgin: Yes, for it cannot be govern'd worse than it is: for the whole
> World is together by the Ears, all up in Wars and Blood, which shews there is
> a general defect in the Rulers and Governors thereof.[48]

Although this subversive suggestion is meant to be humorous—the Virgins are corrected by a Matron who tells them you "talk without knowledge, neither is it fit for such young Ladies as you are to talk of State-matters"—a similar theme appears in another play. In "The Second Part of Bell in Campo," a victorious female commander gains rights by contract from a sovereign; these rights effectively make women citizens of the state—although most of the rights concern the private rather than the public sphere.[49]

In *Blazing World*, Cavendish presents the story of a woman who becomes the absolute ruler of her society after she marries and essentially replaces the emperor of the imagined world. While this might be a form of wish fulfillment for a female writer, the Empress's power does not result in a state free from contention. In fact, the changes she introduces cause faction among her various subjects, which is not resolved until the Soul of the Duchess of Newcastle appears and advises the Empress to restore the former government of the Blazing World, so there will be "one

Soveraign, one Religion, one Law, and one Language, so that all the World might be but as one united Family." The Empress had become too absolute and arbitrary in her rule, but she fears that she will be disgraced if she revokes "her own Decrees, Acts and Laws." To which her natural adviser, the Soul of Cavendish, replies, "That it was so far from a disgrace, as it would rather be for her Majesties eternal honour, to return from a worse to a better, and would express and declare Her to be more than ordinary wise and good."[50] We are once again brought back to the idea that the best form of government is monarchy, as long as the monarch takes advice from wise councillors, in this case Margaret Cavendish (or her soul) herself.

In *Philosophical Letters* (1664), Cavendish claimed that women should not write about politics because "a Woman is not imployed in State Affairs."[51] Her works of the early 1660s belie this claim, whether they consider the composition of the natural or political state, the different kinds of possible governments, or the role of women within or outside the state. Politics are a discussion of who should and who does hold power. Natural philosophy and political philosophy are inextricably linked, as Cavendish recognized in her discussion of Thomas Hobbes that begins *Philosophical Letters.*

Cavendish and Hobbes demonstrate the different routes materialism could take in the mid-seventeenth century, how divergent concepts of matter produced different understandings of animal and human nature, and the extent to which matter and humankind may be described as free either to move or to chose, to rule or be ruled.[52]

## THE MATERIALISM AND POLITICS OF
## CAVENDISH AND HOBBES

Cavendish might have expected that Hobbes would take her seriously because of his close and long-standing relationship to her husband. But the one time they accidentally met, in a London street during her visit to England in the early 1650s, he declined her invitation to dinner. Although they met after her return to England in 1660—Hobbes was living at Chatsworth, home of Newcastle's cousin the Earl of Devonshire and quite close to the duke's residences—there is no evidence that the two ever engaged in serious conversation.[53] Cavendish was forced to confront Hobbes's philosophy in print rather than in person.

Cavendish claimed not to have read the part of *Leviathan* that dealt with politics because, in addition to women having no power in the state, politics was useless unless one had the ear of an absolute king, and, in a statement that would do well in the twenty-first century, politics "requires more Craft then Wisdom."[54]

Much of Cavendish's purported distaste for politics might reflect Newcastle's exclusion from the power center of the new monarchy. The duke had just written a "Letter of Advice" to Charles II, instructing him in the necessary behavior and accomplishments of an absolute prince, which apparently fell on deaf ears.[55] By 1663, Cavendish had already read Hobbes's De Cive (1642), which included an early form of his political theory. Given the relationship between Hobbes and his old patron, Cavendish was certainly aware of his radical political ideas, but probably the last thing she wanted was to be associated with them. The similarity between their two forms of materialism was surely enough.

In Natures Pictures, Cavendish critiqued Hobbes's natural philosophy. Her discussion in "The She-Anchoret" reflects her response to Hobbes's psychological, epistemological, and metaphysical ideas. By 1664, when Philosophical Letters was published, this distancing became even more pronounced, both because of her evolving materialism and perhaps because of the political and religious climate of Restoration England. The new government, by means of the Clarendon Code and the Act of Uniformity, attempted to control liberty of thought and belief. The bishops, meeting in convocation, began a campaign of repression of heterodox ideas, aimed particularly at Hobbes; according to his biographer, the bishops "made a motion to have the good old Gentleman burnt for a heretique."[56] Even Cavendish's husband urged Charles II to censure dangerous ideas, although he had opponents of the monarchy rather than the philosopher in mind.[57]

Ideas are dangerous, and Cavendish knew it. In the preface to Philosophical Letters, she argued that in general, "contradictions are better in general Books, then in particular Families, and in Schools better then in Publick States, and better in Philosophy then in Divinity. All which considered, I shun, as much as I can, not to discourse or write of either Church or State."[58] And in fact, Cavendish proclaimed, "Yea, had I millions of Lives, and every Life was either to suffer torment or live in ease, I would prefer torment for the benefit of the Church; and therefore, if I knew that my Opinions should give any offence to the Church, I should be ready every minute to alter them."[59] The sincerity of this oath of submission is somewhat undercut by the fact that it follows several pages of Cavendish arguing for a reinterpretation of Genesis according to the maxims of her material philosophy. Any orthodox member of the Anglican Church (and any mainstream church) would consider her a blasphemer at best, a heretic at worse. Yet Cavendish took the increasingly unfashionable stand that "opinion is free, and may pass without a pass-port, I took the liberty to declare my own opinions as other Philosophers do."[60] Within books, at least, all ideas could be debated.

In *Philosophical Letters*, Cavendish took the offensive against other natural philosophers. She condemned Hobbesian mechanism, as well as ancient atomism. "I confess," she writes, "there have been many Learned and Studious men, which have been accounted the Sages of Former, Present, and it may be also of Future times; but in my Opinion, they have had very Improbable, and I may say (with-out Disrespect to their Wisdome) very Extravagant Opinions and Phantasms in Natural Philosophy."[61]

One of these extravagant opinions is the mechanical philosophy, in which everything is determined by the pushing of matter in motion: "Another Opinion of some Wise and Learned Men is, that all Exterior Motions, or Local Actions or Accidents proceed from one Motion Pressing upon another, and so one thing Driving and Shoving Another to get each other's Place, which, in my Opinion, if so, no Creature, especially Animals, Could or Should Rest in one Place, but all Creatures in the World would be in a Perpetual Dance, or rather Sliding, which would produce a very Restless Life, and Wearisom to such Lazy Creatures as I am."[62]

Whether laziness is a serious reproach to mechanical determinism, and a reason to reject it, may be open to question. In the 1663 edition of *Philosophical and Physical Opinions*, Cavendish reprinted word for word chapter 59 of the 1655 edition, which was entitled "Of Fortune." This chapter reads like a straightforward endorsement of Hobbesian determinism: "But man, and for all I know, all other things, are governed by outward Objects, they rule and we obey; for we do not rule and they Obey, but everything is led like dogs in a string, by a stronger power [Cavendish's footnote: "Natural power"], but the outward power being invisible, makes us think, we set the rules, and not the outward Causes, for that we are governed by that which is without us, not that which is within us; for man hath no power over himself."[63]

Such physical determinism becomes antithetical to Cavendish's philosophy of free self-moving matter, often repeated in *Philosophical Letters*. Her constantly reiterated theme in the critique of Hobbes in this work is that nature does not move by force: "All the actions of nature are not forced by one part, driving, pressing, or shoving another, as a man doth a wheel-barrow, or a whip a horse; nor by reactions, as if men were at foot-ball or cuffs, or as men with carts meeting each other in a narrow lane."[64]

It is difficult to reconcile a universe governed by the impact of one external object on another with a universe that contains its own internalized freedom, and Cavendish knew it. Part of the explanation for the nearly contemporary publication of these two conflicting ideas may be the fact that Cavendish never wanted to remove anything she had already written from her works. It may be that the ability to see both points of view, which characterized *Orations*, also allowed her to endorse opposing theses in natural philosophy. In fact, later in *Philosophical Letters*, her unease about her own ambiguities is clear. Arguing that absolute free will might diminish the glory of God, she noted, "Wherefore I am neither for Predestination, nor for an Absolute Free-will, neither in Angels, Devils, nor Man; for an absolute Free-will is not competent to any Creature: and though Nature be Infinite, and the Eternal Servant to the Eternal and Infinite God, and can produce Infinite Creatures, yet her Power and Will is not absolute, but limited; that is, she has a natural free-will, but not a supernatural, for she cannot work beyond the power God has given her."[65]

Here is the same problem of reconciling God and freedom Cavendish examined in *Natures Pictures*, but in this instance Cavendish distinguishes between natural and supernatural liberty. In an another discussion of Hobbesian determinism in *Philosophical Letters*, she turned this argument around by insisting that just as God gave nature eternity, he could have given nature and man freedom without undermining his own lordship: but "nevertheless his [man's] motion both of body and mind is free and self-motion, and such a self-motion hath every thing in Nature according to its figure or shape."[66]

Cavendish felt she could escape from the necessitation dilemma by privileging living matter over motion; she saw the preeminence Hobbes gave to motion and in response argued that "Motion is not the cause of Matter, but Matter is the cause of Motion, for Matter might subsist without Motion, but not Motion without Matter."[67] Consequently, action and perception are not caused by the impact of motion on matter but originate in the matter itself. In a long and repetitive discussion, Cavendish argues that Hobbes is wrong to think perception is caused "by some constraint and force." Instead, "all things, and therefore outward objects as well as sensitive organs, have both Sense and Reason, yet neither the objects nor the organs are the cause of them; for Perception is but the effect of the Sensitive and rational Motions, and not the Motions of the Perception." Cavendish's psychology saves her from Hobbesian determinism: "Wherefore, according to my Opinion, the Sensitive and Rational free Motions, do pattern out each others object, as Figure and Voice in each others Eye and Ear; for Life and Knowledge, which I name Rational and Sensitive Matter, are in every Creature, and in all parts of every Creature."[68]

Ontology implies epistemology: self-moving matter knows not because an object impacts it but because it patterns and figures an object in itself, and indeed it is independent of anything but its own subjective reality. Holistic matter vibrates, living and knowing eternally. Not surprisingly, Cavendish rejected the mechanistic notion of inertia because there can never be a state of rest in nature; it is in the nature of all matter to move. and what might seem to be rest is simply a variation in self-motion, caused either internally or externally: "for there is no such thing as Rest in Nature, but there is an alteration of motions and figures in self-moving matter, which alteration causeth variety in opinions, as well as every thing else."[69]

Many of Cavendish's objections to Hobbes's natural philosophy are based on what she considers to be common sense informed by her own philosophy. Thus, she argues, if all perception were caused by pressing, "the sentient . . . would at last be pressed to death." Endeavour, which Hobbes understands as the reactive motion of the sentient body toward a source of motion, would result in pain and ultimately death because "there would, in my opinion, be always a war between the animal senses and the objects, the endeavour of the objects pressing one way, and the senses pressing the other way."[70] Indeed, this struggle of motions that Hobbes envisions would be like "that Custom which formerly hath been used at Newcastle, when a man was married, the guests divided themselves, behind and before the Bridegroom, the one party driving him back, the other forwards, so that one time a Bridegroom was killed in this fashion."[71]

The idea of unruly motions, like unruly guests, causing death and disorder horrified Cavendish. It reeked of the civil war that destroyed the peace of her home country. Although Cavendish disclaimed any ability to comment on political theory, she understood the political implications of a natural philosophy that elevated force and impact into the ultimate causes of the material universe and human psychology. Instead, she argues, "That Nature doth not rule God, nor Man Nature, nor Politick Government Man; for the Effect cannot rule the Cause, but the Cause the Effect." Harmony should be the state of natural and manmade entities, and it cannot be imposed from above but must be the product of the constitutive parts of the whole. Consequently, "Wherefore if men do not naturally agree, Art cannot make unity amongst them, or associate them into one Politick Body and so rule them." In other words, the state constructed by a social contract that imposes a government on man is no more possible than man creating an artificial object distinct from either the body or command of nature. It turns out that both the natural and the political universe is ordered or disordered by Nature, "But man thinks he governs, when as it is Nature that doth it, for as nature doth unite or divide parts regularly or irregularly, and moves the several minds of men and the several parts of mens bodies, so war is made or peace kept."[72]

But nature, in Cavendish's reformulated holistic materialism, is not separate from its parts, which are rather subsumed into its whole. Likewise, the human whole is the human state, which is the result, not the cause, of human association: "Thus it is not the artificial form that governs men in a Politick Government, but a natural power, for though natural motion can make artificial things, yet artificial things cannot make natural power; and we might as well say, nature is governed by the art of nature, as to say man is ruled by the art and invention of men. The truth is, Man rules an artificial Government, and not the Government Man, just like a Watch-maker rules his Watch, and not the Watch the Watch-Maker."[73]

Although this political metaphor sounds eerily like Rousseau's general will, it should not be construed as an endorsement of some form of natural democracy counterposed to Hobbesian absolutism. The integration of parts preempts autonomous actions by any individuated being, although the freedom of self-motion does sometimes result in irregularity and struggle in both nature and humankind:

> I do not say, That man hath an absolute Free-will, or power to move, according to his desire; for it is not conceived that a part can have an absolute power: nevertheless his motion both of body and mind is a free and self-motion, and such a self-motion hath every thing in Nature according to its figure or shape; for motion and figure, being inherent in matter, matter moves figuratively. Yet do I not say, That there is no hindrance, obstruction and opposition in nature; but as there is no particular Creature, that hath an absolute power of self-moving; so that Creature that hath the advantage of strength, subtilty, or policy, shape, or figure and the like, may oppose and over-power another which is inferior to it; yet this hindrance and opposition doth not take away self-motion.[74]

There can be opposition and obstruction in nature and civil war in the state, but these alternative motions do not destroy the self-motion—and unity—of the whole. Indeed, the key to this convoluted passage is the idea that all matter moves figuratively. To understand what Cavendish meant by this, one has to realize the fierce vitalism that characterized her universe. Every part integrated into the whole not only has self-motion; it also has perception, reason, imagination, and passion. To Cavendish, Hobbes is wrong when he argues that "we have no Imagination, whereof we have not formerly had Sense."[75] Every part has the ability to "figure" or create its own notions, ideas, and fancies, which are all corporeal. In a sense, everything makes itself and knows everything through itself. Thus, when Cavendish used the watch/watchmaker analogy elsewhere in her text, she twisted its meaning to empower the parts with an agency compatible with the desires of its designer: "Though the Artificer or Workman be the occasion of the motions of the carved body, yet the motions of the body that is carved, are they which put themselves into such and such a figure, or give themselves such or such a print as the Artificer intended; for a Watch, although the Artist or Watch-maker be the occasional cause that the Watch moves in such or such artificial figure, as the figure or a Watch, yet it is the Watches own motion by which it moves."[76]

The material that makes up the watch is the primary cause of the watch, not the watchmaker, who is only the occasional cause. The mechanists had retained two of Aristotle's four causal categories—the material and the efficient—but Cavendish has eliminated all but the material cause. Whatever form matter takes, it is its own cause and effect, material and being: "There is not any Creature or part of nature without this Life and Soul; and that not onely Animals, but also Vegetables, Minerals and Elements, and what more is in Nature, are endued with this Life and Soul, Sense and Reason."[77]

What distinguishes parts of matter and different forms of matter, according to Cavendish, is only figure and shape. As early as *Poems, and Fancies*, the defining material characteristic of man (always excepting his immortal soul) was the fact that he was upright and bipedal: "But, Matter, you from Figure Forme must take, Different from other creatures, Man must make. For he shall go upright, the rest shall not."[78]

But, according to Hobbes, the distinguishing characteristic of man is that he possesses speech or language, which allows him to ratiocinate by joining the names of things together: "Understanding is nothing else, but Conception caused by Speech. And therefore, if speech be peculiar to man, (as for ought I know it is) then is understanding peculiar to him also."[79] Man also is singular because he has the faculty to foresee the consequences of anything he conceives and he can reason from particulars to universals: "Man hath an other degree of Excellence, that he can by Words reduce the Consequences he finds to General Rules called Theoremes or Aphorisms, that is, he can reason or reckon not onely in Number, but in all other things, whereof one may be added unto, or subtracted from an other."[80] Ultimately, man's ability to ratiocinate will allow him to devise a social contract and escape the state of nature.

Cavendish utterly rejected the idea that the human singularity of speech precludes the ability of other material beings to reason:

> That according to my Reason I cannot perceive, but that all Creatures may do as much [ratiocinate]; but by reason they do it not after the same manner or way as Man, Man denies, they can do it at all; which is very hard; for what man knows, whether Fish do not Know more of the nature of Water, and the ebbing and flowing, and the saltness of the Sea? or whether Birds do not know more of the nature and degrees of Air, or the cause of Tempests? or whether Worms do

not know more or the nature of Earth, and how Plants are produced? or Bees of the several sorts of juices of Flowers, then Men? And whether they do not make there Aphorismes and Theoremes by their manner of Intelligence? For, though they have not the speech of Man, yet thence doth not follow, that they have no Intelligence at all.[81]

If we recall the imaginary realms that populated Cavendish's stories in *Natures Pictures* and look forward to the men-beasts who will live in the Blazing World, it is clear that Cavendish had a sweeping and nontraditional view of humanity and animal-kind. According to Keith Thomas, "In early Modern England the official concept of the animal was a negative one, helping to define by contrast, what was supposedly distinctive and admirable about the human species."[82] Thomas Hobbes certainly reflected this point of view both in his linguistic and political theories: "But the most noble and profitable invention of all other, was that of Speech, consisting of Names or Appelations, and their Connexion; whereby men register their Thoughts; recall them when they are past; and also declare them one to another for mutuall utility and conversation; without which, there had been amongst men, neither Common-wealth, nor Society, nor Contract, nor Peace, no more than amongst Lyons, Bears, and Wolves."[83]

In the most famous line in *Leviathan*, Hobbes contended that in the state of nature, before the establishment of the state, human life was "solitary, poore, nasty, brutish, and short."[84] Natural man is essentially a beast. In the state of nature man shares certain intellectual abilities with the animals. According to Hobbes, understanding "is common to Man and Beast. For a dogge by custome will understand the call, or the rating of his Master; and so will many other Beasts."[85] Humans and beasts both possess the ability to anticipate what will be the consequence of a particular course of action. They are both capable of prudent actions. They are also both capable of understanding the causes of particular effects; according to Hobbes, "The Trayne of regulated thoughts is of two kinds; One, when of an effect imagined, wee seek the causes, or means that produce it: and this is common to Man and Beast."[86] It is the capacity for speech, particularly the ability to name and to understand the connections between names, that elevated humans above animals. The means humans take to overcome their animal fate is the establishment of, in Hobbes's words, "an Artificial Animal," the state, which functions in every way like an automaton and is conceived mechanistically, just as man was in Hobbes's psychological theory. Hobbes's understanding of animals allowed him to construct a revolutionary political theory and to characterize the nature of human rationality and society.

Cavendish rejected Hobbes's view of animal and human rationality completely, and though not specified, this rejection encompassed Hobbes's political theories also. She argued, "But certainly, it is not local motion or speech that makes sense and reason, but sense and reason makes them; neither is sense and reason bound onely to the actions of Man; but it is free to the actions, forms, figures and proprieties of all Creatures; for if none but Man had reason, and none but Animals sense, the World could not be so exact, and so well in order as it is."[87]

All being possesses sense and reason by virtue of the rational and sensitive matter with which it is constituted. Just as the material substratum of nature also contains its own self-regulating principles, beasts possess the capacity to live in harmony and order in a civil society. Since the world of nature is not a chaotic congress of irrational creatures, by implication the state of nature does not consist of bestial human anarchy. Human beings, even before the establishment of the state, are also capable of order and organization, and animals, vegetables, and minerals (and watch parts) are self-regulating and well ordered through their own agency and self-movement.

The all-encompassing internalized freedom of self-movement results in a continuum of spirituality or soul from the smallest piece of matter to the rational soul of man. In fact, when Cavendish explicitly challenged Hobbes in *Philosophical Letters*, her reformulated materialism echoed the traditional great chain of being, where all material being is arranged hierarchically along a continuous ladder. Using the metaphors of the new science, Cavendish described a holistic and diffuse natural world: "nature is one continued Body, for there is no such Vacuum in Nature, . . . nor can any of her Parts subsist single and by it self, but all the Parts of Infinite Nature, although they are in one continued Piece, yet are they several and discerned from each other by their several Figures."[88]

Hobbes had also postulated a resemblance between human and natural motion, but for him this continuity consisted of a similar mechanical reaction to external stimulation. For both Cavendish and Hobbes, human beings are like stones. But for Cavendish, stones live, while for Hobbes, humans act like stones. Hobbes argued, "For every man is desirous of what is good for him, and shuns what is evil, but chiefly the chiefest of natural evils, which is death; and this he does by a certain natural impulsion of nature, no less than that whereby a stone moves downward."[89] In her discussion of Henry More in *Philosophical Letters*, Cavendish argued that stones "have motion, and consequently sense and reason, according to the nature and propriety of their figure, as well as man has according to his."[90]

If all being, from rocks to animals, has life and sense, do women? Chapter 1 showed how animal imagery informs the way Cavendish imagines herself and her productions. Women share the animalism of beasts, but in *Philosophical Letters*, it becomes clear that animals are wise and knowing in their way. Presumably, women can make the same claim. Male ignorance creates the unfounded and arrogant belief that men are superior to other creatures: "But the Ignorance of Man concerning other Creatures is the cause of despising other Creatures, imagining themselves as petty Gods in Nature."[91] Likewise, men, particularly Thomas Hobbes, are wrong when they claim that children do not possess reason.[92] What distinguishes creatures of the same kind is the amount of rational matter they possess. "Nay this difference of the degrees . . . of the rational and sensitive Matter," Cavendish argues, "makes also the difference betwixt particulars in every sort of Creatures, as for example, betwixt several particular Men: But as I said, the nature or essence of the sensitive and rational Matter is the same in all."[93]

An essential equality even within species characterizes material being. Cavendish knew that many men (and women) believed that women were physically and intellectual incapable of masculine activities. One of the characters in her play *Bell in Campo* (1662) declaims, "The Masculine Sex is of an opinion we are only fit to breed and bring forth children" because "our Bodyes seem weak, being delicate and beautifull, and our minds seem fearfull."[94] In *Orations*, a woman speaker laments, "The truth is, we Live like Bats or Owls, Labour like Beasts, and Dye like Worms."[95] Rather than disputing the association of women with beasts, and their innate inferiority, Cavendish developed a natural philosophy that undermined the differences between the various parts of nature. An identity of matter substantiated an equality of being, even if the necessity of order in nature and the state necessitated that some command and others obey.

As we saw above, Cavendish argued in *Sociable Letters* that women cannot be political because they are neither subjects of an absolute state nor citizens of a commonwealth. There would be no place for them either in the England of Charles II or the Leviathan of Thomas Hobbes. Unlike men, who cause disorder in the state, women, at least during the civil war, are peaceful: in *Sociable Letters*, Cavendish writes her imagined correspondent, "But however, Madam, the disturbance in this Countrey hath made no breach of Friendship betwixt us, for though there hath been a Civil War in the Kingdom, and a general War amongst the Men, yet there hath been none amongst the Women."[96]

Women do not make war, nor do beasts, with whom Cavendish is happy to be associated, both metaphorically and scientifically:

Perchance some may say, that if my Understanding be most of Sheep and a Grange, it is a Beastly Understanding: My answer is, I wish Men were as

Harmless as most Beasts are, then surely the World would be more Quiet and Happy than it is, for then there would not be such Pride, Vanity, Ambition, Covetousness, Faction, Treachery, and Treason, as is now. Indeed, one might very well say in his Prayers to God, O Lord God, I do beseech thee of thy infinite Mercy, to make Man so, and order his Mind, Thoughts, Passions, and Appetites, like Beasts, that they may be Temperate, Sociable, Laborious, Patient, Prudent, Provident, Brotherly-loving and Neighborly-kind, all which Beasts are, but most Men not.[97]

If only men were more like beasts, and presumably more like women, all the wars and disorders that plague the world could be avoided. However, Cavendish never overtly endorsed this conclusion. She was well aware that there were many women who act out of passion and cause discord within the "matrimonial Government," and even participated in the civil war, but she also thought that this could be avoided if women were "inclosed with Study, Instructed by Learning, and Governed by Knowledg and Understanding."[98] In fact, it is also true that young men cause "Combustions and Disorders . . . but it is against Nature for Young men to be Wise, wherefore they are fitter to Obey than to Command."[99]

All creatures in Cavendish's universe are capable of learning and knowledge. If the innate rationality of all were recognized, presumably their worth would also be acknowledged and respected. Cavendish's critique of Hobbes's materialism and psychology, and ultimately his political philosophy, denied that nature is governed by force, directing how one either acts or perceives. In order to eliminate constraint from her natural philosophy, Cavendish developed a theory of matter independent of coercion from external sources. Her theory of holistic matter, like Hobbes's mechanistic materialism, had implications far beyond the natural world. It allowed her to envision a commonwealth in which the king was not a tyrant but the benevolent ruler of his people. It allowed her to imagine a world in which peace rather than strife would be the natural state of humanity and nature, in which animals would be the model of community rather than anarchy. This world—and the difficulties in producing it—would be described in even more detail when Cavendish returned once more to fantasy in *Blazing World* in 1666.

As Cavendish's materialism evolved, so did her political ideas. Although she wrote that women were outside the political realm, she found that becoming a citizen of the world of letters required her to become politically engaged: natural philosophy was neither removed from nor without implications for political philosophy. The constituents of matter—rational, sensitive, and inanimate—live in civil accord with one another just as she hoped the bodies they constituted live in harmony. Matter blended into one substance, animals into peaceful unity, and people into an ordered polity. While Cavendish's early works emphasized the discord caused by motion in a material world, carving antagonistic parts out of matter, her later works incorporated matter and motion into a holistic and vital materialism, substantiating at the imperceptible level the order of a well-functioning state. Implicit in this argument is the idea that all being, including women and beasts and stones, possesses rationality and freedom, bringing its powers into natural and political associations.

# chapter six

# The Challenge of
# Immaterial Matter

*I*n the mid-1660s, Cavendish's feelings of alienation from both the civil sphere and the republic of letters increased dramatically. Her early works had not received the response for which she had hoped. She thought she "should be highly applauded" because "self-conceit, which is natural to mankind, especially to our Sex, did flatter and secretly perswade me, that my Writings had Sense and Reason, Wit and Variety."[1] She lamented in *Sociable Letters* that she could not expound her philosophy in public, like male philosophers; "I fear," she writes, "the Right Understanding of my Philosophical Opinions are likely to be Lost, for want of a Right Explanation, for they may be Interpreted not the way I Conceiv'd them, that is, not to my Sense or Meaning."[2]

In order to receive the attention her works warranted, and explain their contents, Cavendish launched a kind of assault against the most famous philosophers of her age, attacking their works in print and contrasting their ideas with her own. She was careful to indicate her respect for them: "I am confident there is not any body, that doth esteem, respect and honour learned and ingenious Persons more then I do," but that did not stop her from criticizing their works in terms of her own philosophy.[3]

Cavendish wanted to cross borders into a masculine country and confront the natives to validate her position among them. In *Philosophical Letters*, Cavendish picked and chose among those parts of the writings of other philosophers that interested her. Her critiques of their philosophies reveal what she considered most important in natural philosophy. Not surprisingly, matter theory and motion are her paramount concerns, and she reiterates her holistic materialism many times—even stopping in the middle of her critiques of other philosophers to assert her own system, over and over again. How much she really understood of their thought is open to question, but she clearly understood where their ideas contradicted her own, and she was willing to defend herself and criticize them.

Cavendish attacked the materialist Hobbes in *Philosophical Letters*, but most of this work critiques the most prominent immaterialists of her age, those who postulated some kind of spiritual, nonmaterial force or being in the universe that operated on matter. She criticized René Descartes for separating motion and matter and for his dualistic metaphysics. Not surprisingly, his doctrine of animal automata appalled her. Likewise, she attacked Henry More's doctrine of immaterial spirits and the way he subjugated a female anthropomorphized Nature to an overriding Spirit of Nature. She found Joan Baptista Van Helmont's obscure and impious mixture of spirit and matter particularly appalling and traced the way his ideas could endanger women and even reshape the story of the female figures in the Christian account of creation and salvation.

Cavendish wanted the male thinkers to take her philosophic ideas as seriously as she took theirs. She argued that her discourse should not be limited to women: "I have been informed, that if I should be answered in my Writings, it would be done rather under the name and cover of a Woman, then of a Man, the Reason is, because no man dare or will set his name to the contradiction of a Lady." Indeed, a man had already written against some sections of *The Worlds Olio*, but under a female pseudonym. Such "a Hermaphroditical Book," Cavendish argued, is "not worthy taking notice of."[4] To be a philosopher or a "philosopheress" takes long study, but the productions of both sexes can be treated with mutual respect. "But I cannot conceive why it should be a disgrace to any man to maintain his own or others opinions against a woman," insisted Cavendish, "so it be done with respect and civility."[5]

Rather, she suggested, her ability to debate ideas with men is a basic point of freedom. Cavendish argued that since "in Natural Philosophy, Opinions have Freedome, I hope these my Opinions may also Injoy the same Liberty and Privilege that others have, which without great Injustice no Body can Deny me."[6] The scientific world, according to Cavendish, should be characterized by a free flow of ideas, articulated by individuals with the freedom to do what they want. In the seventeenth century, lack of restraint, the ability to do what one wanted, was an essential characteristic of the notion of freedom and a privilege associated with free men and gentlemen. Freedom was a mark of status and a mark of power. In arguing that her ideas should be accorded the same hearing as others, Cavendish claimed the honor, status, freedom, and power of her male counterparts for both herself and her sex.

The concern with, and constraint of, gender is more evident in *Philosophical Letters* than in any of Cavendish's other natural philosophic texts. Still, as she takes on the greatest thinkers of her time, her demand for philosophic equality sometimes seems to be undermined by self-doubt. She justifies any errors by pleading her sex: "I considered first, that those Worthy Authors, were they my censurers, would not deny me the same liberty they take themselves; which is, that I may dissent from their Opinions, as well as they dissent from others, and from amongst themselves: And if I should express more Vanity then Wit, more Ignorance then Knowledge, more Folly then Discretion, it being accorded to the Nature of our Sex, I hoped that my Masculine Readers would civilly excuse me, and my Female Readers could not justly condemn me."[7] However, she adds, her lack of wit and sagacity in her earlier works were the results of her lack of exposure to the "Terms or Expressions of Natural Philosophy" and rather "did merely issue from the Fountain of my own Brain." Cavendish now will make up for those shortcomings and follow the commands of her anonymous female correspondent, Madam, in confronting the ideas of others and comparing them with her own.

At the command of a woman—that is, herself—Cavendish will overcome her former defects of femininity by clarifying her ideas for an audience both male and female. Thus, while much of the material in the *Philosophical Letters* seems to be gender-neutral, the book is framed in gender terms. This is true of her critique of Hobbes and her disagreements with the three immaterialist philosophers she attacks in the rest of the book. In fact, Cavendish argues that in an inversion of traditional gender characteristics, Descartes, Henry More, and Van Helmont are guilty of the same kinds of flights of fancy she had displayed in her earlier works, particularly when they try to harness God in their own imagined systems: "But some Philosophers striving to express their wit, obstruct reason; and drawing Divinity to prove Sense and Reason, weaken Faith so, as their Mixed Divine Philosophy becomes meer Poetical Fictions, and Romancical expressions, making material Bodies immaterial Spirits, and immaterial Spirits material Bodies; and some have conceived some things neither to be Material nor Immaterial, but between both."[8]

In short, the philosophers have jumbled the natural and divine. If God and "his servant Nature" were as mixed up as the philosophers are in their descriptions, "the Universe and Production of all Creatures would soon be without Order and Government, so as there would be a horrid and Eternal War both in Heaven, and in the World."[9] Philosophic systems have theological, natural, and civil implications. The philosophers with their words create a pattern of confusion: their fancies and romances undermine the harmony of the universe—a harmony revealed in Margaret Cavendish's natural philosophy, which separates the divine and the natural, supernatural spirit and matter, God and Nature.

## DESCARTES

Some scholars link René Descartes with feminism, or at least the possibility of the ungendered mind, while others think that his philosophy provided yet another assault on women by denigrating the body with which women were associated. Arguing for the appeal of Cartesianism to women, Erica Harth writes, "A major source of Descartes's initial appeal was the apparent universality of his message that his rules and method for discovering truth could be used by anyone, of either sex." Thus, "his dualist conception of mind and body strengthened the Augustinian concept of mind as a place 'where there is no sex.'"[10]

There may have been Cartesian women in the seventeenth century, but Margaret Cavendish was not among them. Cavendish's own theory of matter—living, self-moving, and self-conscious—obviates the necessity for an immaterial soul and hence refutes the Cartesian duality of matter and mind. Her epistemology, focused on the subjective cognition of animated matter, was the opposite of the Cartesian philosophy, with its elevation of mind as the only possessor of rationality. Cavendish repeatedly denied the possibility of unextended substance as separate from, and superior to, extended matter: "If he mean the natural mind and soul of Man, not the supernatural or divine, I am far from his opinion; for though the mind moveth onely in its own parts, and not upon, or with the parts of inanimate matter, yet it cannot be separated from these parts of matter, and subsist by it self, as being part of one and the same matter the inanimate is of, (for there is but one onely matter, and one kind of matter, although of several degrees) onely it is the self-moving part."[11]

Animate matter, according to Cavendish, is composed partially of rational matter—which, in her critique of Descartes, Cavendish figures as female.[12] It is very unusual for her to expressly gender her matter theory, and her doing so here may reveal how antithetical Cavendish believed the Cartesian separation of body and soul to be to her own philosophy. Although she often uses the terms *nature* and *matter* interchangeably, nature is gendered as female only when Cavendish personifies it as the externalized self-governing principle of the totality of self-moving matter, rather than the inherent life and motion of matter.[13] While this might seem a distinction without a difference, Cavendish did mean for her material substance to transcend gendered meanings—except when she was discussing Descartes.

But whether any materialist could entirely escape her or his culture-bound ways of thinking about nature is certainly open to question. At least in Cavendish's personified depiction of nature, nature remained sentient and female in the writer's natural philosophy, which was thus anti-Cartesian. Modern feminist scholars have accused Descartes of instituting a male monopoly of philosophy that, according to Joanna Hodge, "has served both to exclude women from philosophy and to obscure how that exclusion has taken place."[14] Karl Stern argues that one encounters in Cartesian rationalism "a masculinization of thought," whereby "the rationalist, whether male or female, is in retreat from those aspects of experience culturally associated with the maternal and the feminine."[15] Other recent critics emphasize the epistemological implications of Cartesianism, which separates the object of knowing from the knowing subject, with reason being the masculine knower and body the defeminized and passive other.

But Cavendish's critique did not directly address the question of whether Descartes's philosophy had positive or negative implications for women. Rather, she emphasized the mistake Descartes made in separating motion from body, thereby draining matter of any vitality and precluding any kind of animal rationality. She also found Descartes's rationalist epistemology repugnant. What Harth calls the "universalizing, sex-neutral rational discourse" had no appeal for Cavendish.[16] But most of Cavendish's criticism concerns not the metaphysical but the natural philosophic elements of his system. She thought his notion of material vortices was downright silly and his attempt to connect body and mind through the pineal gland ridiculous. Unlike some of her female contemporaries, such as Anne Conway (1630–79) and Princess Elisabeth of the Palatinate (1630–1714), who saw much to value in Descartes's philosophy, Cavendish attacked the particulars of his theories, and her critique sometimes displays a gendered perspective.[17]

Cavendish had a long acquaintance with the ideas of Descartes, although she read his work only after she had some parts of his *Principia*, *Dioptics* and *Discourse on Reason* translated into English before writing *Philosophical Letters*. Her section on Descartes is the shortest discussion of her four principal authors in *Philosophical Letters*: it is less than 40 pages, unlike the nearly 100 each she devoted to Hobbes and Henry More and the nearly 200 she gave to Van Helmont. Her inability to read French and Latin made most of Descartes's writings unavailable to her, although her knowledge of the French philosopher also drew on conversations with her husband and brother-in-law. Both Newcastle and Sir Charles Cavendish were acquaintances and correspondents of the French thinker. According to Descartes's seventeenth-century biographer, Adrien Baillet, Sir Charles was "was desperately in love with Monsieur Des Cartes his Philosophy" and sought to bring him to the court of Charles I in 1640. The scheme was blocked only by Descartes's concern that the king would not be able to support natural philosophy because of the civil war.[18] Sir Charles was most interested in Descartes's mathematical ideas, which he discussed in a series of letters to the mathematician John Pell during the 1640s. In 1645 and 1646, Descartes sent two important letters to Newcastle about the Frenchman's theories of sensation and animal automata. The Cavendish brothers finally met Descartes in Paris in 1647, when he dined at their home, along with Thomas Hobbes and Pierre Gassendi, in an effort to reconcile the three most important mechanistic philosophers of the early seventeenth century. Margaret Cavendish probably met Descartes at this dinner.

Sir Charles had long tried to mediate the French philosopher's ongoing dispute with Hobbes about issues of priority, metaphysics, and mathematics. He followed Descartes's publications with great interest and enthusiasm. By 1650, he had read the "excellent man's" *Principia, Dioptics* and *Passions of the Soul.* It is impossible that with her lively interest in things philosophic, Cavendish was unaware of Descartes's philosophy. In fact, Sir Charles may be implicitly referring to her when he wrote to Pell in March 1646, "I am most glad that you are entered in to an acquaintance with M[r]: de Cartes, who doubtless is an excellent man, yet let not his discourse perswade you from writing and printing your intended workes, for I conceive not but that the greatest witts and most learned doe much benefit on an other by publishing and communicating theyre choice thoughts one to an other, and that even meaner witts and Clerkes gleane something from them too."[19]

Margaret Cavendish had plenty of opportunity to benefit from Sir Charles's optimistic view of the potential of even "meaner wits" to understand something about natural philosophy. She and Sir Charles traveled to England in 1651, and some hints of their conversations appear in *Poems, and Fancies.* Clearly their discussions concerned atomism, but there is also a possible reference to Descartes in the description of fairyland. As we saw in Chapter 2, Cavendish conceived of fairies as reified atoms. The home of the king of these atomic fairies seems to be a Cartesian brain. According to Cavendish's description of "The City of the Fairies,"

> The City is the Braine, incompast in
> Double walls (*Dura Mater, Pia Mater* thin)
> It's trenched round about with a thick scull,
> And fac'd without with wondrous Art, and skill. . . .
> And Oberon King dwels never any where,
> But in a Royall Head, whose court is there:
> Which is the kernell of the Braine,
> We there might view him, and his beauteous Queen.[20]

In the seventeenth century, a kernel was a very specific part of the brain. It was the pineal gland, and in the Cartesian system, it is the place where mind and body interact.[21] It is interesting that a male lives in this place—a male, moreover, who parallels the often absent God of Cavendish's early natural philosophy and anticipates the Emperor of the Blazing World, who gives up his authority to the Empress, Cavendish's other self. In fairyland, "There Mab is queen of all, by Natures will, / And by her favour she doth governe still."[22] It may be that an impotent king, confined within a gland in the brain, is an avatar for an equally useless philosopher.

Years later, after Cavendish had read the parts of Descartes's *Principia* concentrating on the mind-body problem (which had been translated for her), the question of the kernel of the brain pops up again. In *Philosophical Letters*, she writes, "Neither can I apprehend, that the Mind's or Soul's seat should be in the Glandula or kernel of Brain, and there sit, like a Spider in a Cobweb, to whom the least motion of the Cobweb gives intelligence of a Flye. . . . And that the Brain should get intelligence by the animal spirits as his servants, which run to and fro like Ants to inform it."[23] In fact, Cavendish probably knew about Descartes's theory of perception long before she read about it. Descartes had described it to Newcastle in a letter written in October 1645: "I am convinced that hunger and thirst are felt in the same manner as colours, sounds, smells, and in general all the objects of the external senses, that is, by means of nerves stretched like fine threads from the brain to all other parts of the body. They are so disposed that whenever one of these parts is moved, the place in the brain where the nerves originate [the pineal gland] moves also, and its movement arouses in the soul the sensation attributed to the rest."[24]

For Descartes, a mechanistic account of sensation supported his theory of the utter separation of mind and body: one is a rational substance, and the other is extended substance. By 1664, Cavendish believed she had the ammunition she needed to demolish Cartesian dualism: her own holistic materialism. She argues, "We cannot assign a certain seat or place to the rational, another to the sensitive, and another to the inanimate, but they are diffused and intermixt throughout all the body." In addition, she denies Descartes's claim that mind "is a substance distinct from body, and may be actually separated from it and subsist without it." While there is a kind of matter that can be called rational, it is coextensive and fused with the sensitive and inanimate matter.[25] Holistic matter incorporates animate and inanimate matter—there is no place or necessity for the separation of mind and body.

Cavendish does not disagree with Descartes's argument that all matter is itself composed of the same substance and that these parts are moving.[26] What she does condemn is his separation of motion from body, which she quotes from the *Prin-*

*cipia*: "Motion is onely a mode of a thing, and not the thing or body it selfe." Basic to her own system is the claim that "Motion is but one thing with body, without separation or abstraction soever."[27] And for Cavendish, wherever there is moving body, there is life and perception. Consequently, Descartes's argument that corporeal bodies, including the human body, act in a mechanistic fashion must be wrong. Citing her own profoundly subjective psychology, Cavendish argues that each particle moves and knows through itself: the watch, not the watchmaker, moves its own parts. Cognition and perception, which Descartes separates into distinct functions of the rational soul and extended substance, are collapsed in Cavendish's formulation. Whereas Descartes had argued that rational thought occurred only in the immaterial mind, Cavendish reversed the process and argued that rationality itself is a part of matter.

Cavendish's objection to Cartesian dualism emanated from her profound materialism, which also led her to condemn other aspects of his thought that contradicted her own. One of the doctrines she most vehemently attacked was Descartes's theory of animal automatonism. Descartes argued that just as the corporal parts of the human body act like a machine, animals too lack rationality and indeed are no more than moving machines. The sensations and passions animals experience are simply due to the actions of the animal spirits on the nerves and do not demonstrate consciousness.[28] Descartes included this doctrine in several of his early works, including *Discourse on Method* (1637), parts of which Cavendish had translated for her use in 1664. Descartes also defended this idea to Newcastle, who presumably had challenged the assertion that animals, including the horses Newcastle loved so much, lacked rationality. Descartes wrote to him in 1646, "All the things which dogs, horses, and monkeys are taught to perform are only the expressions of their fear, their hope, and their joy; and consequently they can be performed without any thought." Animals do not have language, unlike even a deaf and dumb human being, who can express his or her thoughts with signs; animal sounds are simply noise without meaning. "I know that animals do many things better than we do, but this does not surprise me," Descartes explained. "It can even be used to prove they act naturally and mechanically, like a clock, which tells the time better than our judgement does."[29]

But for Cavendish, if even mechanical clocks have their own kind of understanding, surely there is no such thing as an animal machine. In *Philosophical Letters*, she argues that just as each part of the body has its own kind of knowing, and just as each man has his own kind of knowledge, likewise all animals have their own kind of knowledge, even if they cannot speak to one another. Cavendish repeats an argument she had used against Hobbes, but for a different purpose.

Animals possess reason, she explains, "for reason is the rational part of matter, and makes perception, observation, and intelligence different in every creature, and every sort of creatures, according to their proper natures. . . . Wherefore though other Creatures have not the speech, nor Mathematical rules and demonstrations, with other Arts and Sciences, as Men; yet may their perceptions and observations be as wise as Men's, and they may have as much intelligence and commerce betwixt each other, after their own manner and way, as men have after theirs." Cavendish had criticized Hobbes for denying animals free will; here she criticizes Descartes for denying other creatures' intelligence and rationality. Cavendish recognized that Descartes's doctrine was ultimately an argument about power, and the power of one species over another, based on man's "conceited prerogative and excellence."[30]

In Descartes's second letter to Newcastle, he argued about animals "that if they thought as we do, they would have an immortal soul like us. This is unlikely, because there is no reason to believe it of some animals without believing it of all, and many of them such as oysters and sponges are too imperfect for this to be credible."[31] Cavendish leaves the question of the immortal soul out of her philosophy; like the nature of God, it is impossible for humans to understand.[32] But when it is an issue of the mind, and its ability to think, there are no things in nature that lack cognition. There are no creatures so imperfect or so inert that they do not possess some kind of life and rationality. So, in the discussion that concludes her examination of Descartes, she adds, "I have seen pumpt out of a water pump small worms which could hardly be discerned but by a bright Sun-light . . . yet they were more agil and fuler of life, then many a creature of a bigger size." Indeed, there are many creatures even smaller, or even greater, that the senses cannot perceive, but "sense and reason inform us, that there are different degrees in Purity and Rarity, so also in shapes, figures and sizes in all natural creatures."[33] Cavendish is attacking not only the Cartesian animal automata here but also Cartesian epistemology. Reason alone cannot reveal the attributes of nature; instead it must include sense—both the rational and sensitive matter enable human understanding. Descartes's epistemology is wrong, she argues: "It is not onely the Mind that perceives in the kernel of the Brain, . . . there is a double perception, rational and sensitive, and . . . the mind perceives by the rational, but the body and the sensitive organs by the sensitive perception; and as there is a double perception so there is also a double knowledge, rational and sensitive, one belonging to the mind, the other to the body."[34]

Moreover, Cavendish argues, Descartes is incorrect to maintain that the ontological properties of nature, revealed by the rational mind, are separate from matter itself. Place and time are adjuncts of matter and do not exist separately from it.[35] Motion itself cannot be transferred by impact, as Descartes argues, because it is not a separable adjunct to matter.[36] Extended substance contains not only primary qualities, Cavendish adds, but secondary ones. Matter's motion is not necessarily circular as Descartes had argued; nor is the world of matter in motion made up of vortices—or, as Cavendish calls them, "aethereal Whirlpooles"—composed of three different sizes of matter. The last claims are, in fact, "rather Fancies, then rationall or probable conceptions," for, she asks, "how can we imagine that the Universe was set a moving as a Top by a Whip, or a Wheele by the hand of a Spinster?"[37] These analogies are interesting; apparently Cavendish thought the Cartesian universe was either childish or feminine. The association of nature with the female and irrational in this instance becomes possible only in a world denuded of self-activity.

Of course, Cavendish's own universe was guided by a female principle, but it is a principle inherent in matter itself. Nature "hath an Infinite natural power, yet she doth not put this power in act in her particulars; and although she has an infinite force or strength, yet she doth not use this force of strength in her parts."[38] Nature is a self-regulating presence that does not compel or force her own parts. Unlike Descartes's artificial universe, created and moved by an external artificer, infinitely moving matter regulates itself.

Cavendish's critiques of Descartes, whether of his dualism of mind and body, of his privileging of the immortal and rational soul over inert body, or of the details of his physics, physiology, and psychology, are ultimately based on the living nature of Nature and matter. Self-conscious matter obviates the need for a separate natural soul. Self-moving matter is living matter. Self-knowing matter endows all creatures with cognizance, at least of their own kind. To the extent that nature and matter are gendered, so is Cavendish's response to Descartes. Thus, unlike some of her female contemporaries who were seen as sympathetic to the French philosopher, we might call her an anti-Cartesian woman.

## HENRY MORE

It is perhaps pointless for historians to play with counterfactuals, but if Henry More had read Margaret Cavendish's work, he would have been horrified. The English Cambridge Platonist was a fierce opponent of the materialism of Thomas Hobbes and, after a period of infatuation, the dualism of Descartes.[39] In 1663, Cavendish provided him with the opportunity to react to her own version of materialism when she sent him copies of her work. More's thank-you note expressed his astonishment at receiving this gift but also the fact that he had not sat down and perused her volumes:

> I was very much surprized when your Servant saluted me from so Illustrious a personage; but when he produced those noble Volumes as an intended Testimony of your Ladyships respect, the unexpectedness of so great an honour made me suspect the Messenger of a mistake. . . . Madam, I humbly crave your Pardon for my boldness, and impatience that I offer so hastily to return thanks for so eminent a Favour, before I have computed the value thereof, nor as yet fitly polished and adorned my Stile, by a longer converse with your Ladyships most Elegant and Ingenious Writings.[40]

There is no evidence that More ever actually took the time to read Cavendish's natural philosophy. At least, it is clear that any good will she might have felt toward him evaporated by the next year when, reacting to his lack of response, she included him in the group of natural philosophers she criticized in *Philosophical Letters*.

Cavendish had challenged Henry More to treat her like a peer, and she approached him as an equal. When she examined his philosophy, she found little to like. Her critique was based on a fairly close reading of More's *Antidote to Atheisme* (1653) and *Immortality of the Soul* (1659). In these works, Cavendish discovered a system as unique, and perhaps as odd, as her own.[41] Cavendish and More agreed that it was impossible to explain the nature of the universe on the basis of mechanistic principles alone, and both thought that Descartes's unextended substance failed to provide an explanation of material action. As we know, Cavendish supplied this lack with her theory of self-moving, self-conscious matter, but More took the opposite course. He argued that matter was inert and passive and that the harmonious action of the world could be explained only by what he calls "the Hylarchic Principle." According to this idea, there is in nature "a substance incorporeal, but without sense and animadversion, pervading the whole matter of the universe, and exercising a plastic power therein . . . , raising such phenomena in the world, by directing the parts of the matter, and their motion, as cannot be resolved into mere mechanical powers."[42] This extended incorporeal substance is the agent of God and functions as a kind of noncognizant lieutenant who orders the world according to God's will. More's Spirit of Nature, therefore, obviates the need for Nature herself, whether as an external organizing principle or as a force inherent in matter itself.

Cavendish's attack on Henry More was about sex and power. She rejected his philosophy not only because it was a version of immaterialism but also because it replaced a feminine force with a masculine agent. The inherently gendered character of Cavendish's natural philosophy is nowhere so apparent as in her critique of the Cambridge Platonist.

Cavendish began by denying the possibility of an immaterial and extended substance: "He thinks it is absurd to say, that the World is composed of meer self-moving Matter, [but] it is more absurd to believe Immaterial substances or spirits in Nature, as also a spirit of Nature, which is the Vicarious power of God upon Matter." The absurdity of this belief, moreover, is not just that it is impossible to accept the existence of an extended immaterial substance (which it is) but that such a belief diminishes the power of God and the power of Nature. Cavendish writes, "For why should it not be as probable, that God did give Matter a self-moving power to her self, as to have made another Creature to govern her: For Nature is not a Babe, or Child, to need such a Spiritual Nurse, to teach her to go, or to move; neither is she so young a Lady as to have need of a Governess, for surely she can govern her self; she needs not a Guardian for fear she should run away with a younger Brother, or one that cannot make her a Jointure."[43]

As we saw in Chapter 5, in *Sociable Letters* Cavendish had claimed that women had no power in the state and indeed were neither citizens of the state nor its subjects.[44] In this, she emphasized the de jure political powerlessness of early modern women, at least in England. According to Merry Wiesner, "In England, a married woman was not even considered a legal person under the common law, but was totally subsumed within the legal identity of her husband."[45] Roman law also denied women legal rights owing to their "fragility, imbecility, irresponsibility, and ignorance," and the result was that with Roman law's reintroduction in early modern Europe, "unmarried adult women and widows were again given male guardians, and prohibited from making any financial decisions, even donations to religious institutions, without their approval."[46]

But the realities of government and commerce necessitated that in some cases women could have an independent legal personality and the right to sign contracts, inherit property, and even appear in court. Such rights were particularly important for single women and widows who often served as guardians for their minor children. However, in England after the Restoration, single women and widows lost rights they had possessed since the later Middle Ages, even though they continued to have civil responsibilities, such as paying taxes and maintaining their property. When civil rights and societal norms clashed, according to Mendelson and Crawford, "a consensus had emerged which declared women unfit for civic office, and which rated the preservation of gender order as a higher propriety than women's proprietary rights."[47]

Married women in particular were subsumed and subordinated in the patriarchal household. Cavendish experienced the extent that married women were dependent on male legal authority early in her marriage. We have seen that the one time Cavendish attempted to appear before a committee of Parliament, to argue for a portion of her husband's sequestered estate, the affair ended in embarrassment and failure. "I whisperingly spoke to my brother to conduct me out of that ungentlemanly place, so without speaking to them one word good or bad, I returned to my Lodging, & as that Committee was the first, so it was the last, I ever was at as a Petitioner." In fact, in 1653, when Cavendish described this event, she condemned the new laws enacted during the Commonwealth, "where Women become Pleaders, Attorneys, Petitioners and the like." This statement speaks to Cavendish's ambivalence about female capacity, but it also may indicate her discomfort at the subservience any woman had to show before a higher authority.[48]

Cavendish was well aware that some women could and did exercise authority in their own households and even the world at large. She had served at the court of Queen Henrietta Maria, who played both a political and military role during the civil war and exile.[49] Cavendish's own mother was a strong and independent woman, who as a widow successfully fought for the guardianship of her minor son and ran her own household firmly and efficiently.[50] In 1656, Cavendish described her mother as possessing "a Magestick Grandeur," a woman who, while claiming she was too weak to manage her family and household, nevertheless "took a pleasure and some little pride in the governing thereof; she was very skilfull in Leases, and setting of Lands, and Court-keeping [presiding at manorial courts], ordering of Stewards, and the like affaires."[51]

Clearly, as Cavendish wrote her natural philosophy, she developed a philosophy that reflected the actual autonomy one woman—Nature—had from the dominance of the male. More's Hylarchic Principle prompted Cavendish to clarify her understanding of the relationship between God and Nature. While More's Spirit of Nature was essentially a cipher of God's authority, Nature functioned as her own woman. Cavendish argues,

> Some of our modern Philosophers think they do God good service, when they endeavour to prove Nature, as Gods good Servant, to be stupid, ignorant, foolish and mad, or any thing other then wise, and yet they believe themselves wise, as if they were no part of Nature; but I cannot imagine any reason why they should rail on her, except Nature had not given them as a great a share or portion, as she hath given to others. . . . However, Nature can do more then any of her Creatures: and if Man can Paint, Imbroider, Carve, Ingrave curiously,

why may not Nature have more Ingenuity, Wit and Wisdom then any of her particular Creatures? The same may be said of her Government.[52]

Nature is not a stupid servant of God, as envious men may believe, or his wife. Rather, when God created moving matter, he also gave Nature self-governance. Nature has power over herself—although not an absolute power, which belongs only to God.[53] To More's contention that if matter were self-moving it would move away from the blows of a hammer or the heat of a fire and therefore it must be passive, Cavendish argues that while some parts of nature can be overpowered by force, power itself remains in Nature, "for not every particular creature hath an absolute power, the power being in the Infinite whole, and not in single divided parts." In fact, More's problem in understanding the nature of Nature is that he can think only in a masculine way: "Man thinks Nature's wise, subtil and lively actions, are as his own gross actions, conceiving them to be constrained and turbulent, not free and easie, as well as wise and knowing; Whereas Nature's Creating, Generating and Producing actions are but an easie connexion of parts to parts, without Counterbuffs, Joggs and Jolts, producing a particular figure by degrees, and in order and method, as human sense and reason may well perceive."[54]

Sometimes it is a mistake to find a transparent gendered meaning in metaphorical language, but it is justified here. Men cannot conceive of nature in anything but their own image, according to Cavendish: "Some in their opinions do conceive Nature according to the measure of themselves." Their lack of imagination results in a basic misunderstanding of the world: actions in nature are not forced but free; Nature does not push or shove, she creates and generates. Imagining an immaterial spirit as a necessary ordering device misconstrues how nature works and derogates the glory of God, "for all Nature's free power of moving and wisdom is a gift of God, and proceeds from him; but I must confess, it destroys the power of Immaterial Substances, for Nature will not be ruled and governed by them."[55] Indeed, to extend Nature so she is composed of both material and immaterial parts, Cavendish argues, attenuates and destroys nature: if there were a proper understanding of Nature, "there would not be so many strange opinions concerning nature and her actions, making the purest and subtillest part of matter immaterial or incorporeal, which is as much, as to extend her beyond nature, and to rack her quite to nothing."[56]

The historian Carolyn Merchant argues that the new natural philosophers of the seventeenth century conceived of nature as an unruly woman who has to be tortured into submission.[57] Thus, for example, Francis Bacon wrote, "Neither ought a man to make scruple of entering and penetrating into these wholes and corners [of nature], when the inquisition of truth is his whole object."[58] Did Cavendish envision her male counterparts as part of a conspiracy to rack and rape nature? Did she see their philosophies of nature as the imposition of male power on passive matter, whether in the Cartesian form of the dominance of rational unextended substance or in Henry More's determining immaterial spirit? She did directly criticize More's statement that "Immaterial substance, Indivisible, that can move it self, can penetrate, contract and dilate it self, and can also move and alter matter." She argued instead, "For Matter by the Power of God is self-moving, and all sorts of motions, as contraction, dilation, alteration, penetration, etc. do properly belong to Matter; so that natural Matter stands in no need to have some Immaterial or Incorporeal substance to move, guide and govern her, but she is able enough to do it all her self, by the free Gift of the Omnipotent God."[59]

The essence of Cavendish's argument against More is that it is unnecessary to invent another explanation to understand nature's motions or the natural order and harmony of the world. In fact, such an argument invites paganism: "But I fear," she remarks, "the opinion of Immaterial substances in Nature will at last bring in again the Heathen Religion, and make us believe a god Pan, Bacchus, Ceres, Venus, and the like, so we may become worshippers of Groves and shadows, Beans and Onions, as our Forefathers."[60]

Cavendish knew that her own philosophy was open to the charge of making Nature into a goddess. As she had done many times before, Cavendish protested that when she made Nature eternal, she did not challenge the supremacy of God. Indeed, she even cites Genesis to prove her contention that an eternal nature is the product of an eternal God, although she soon retreats to her fideistic position that it is impossible for humans to understand how this may be.[61] She advises More that theological questions, such as the nature of the immortal soul and the nature of God himself, should be left to the church: "But if man had a mind to shew Learning, and exercise his Wit, certainly there are other subjects, wherein he can do it with more profit, and less danger, then by proving Christian religion by Natural Philosophy, which is the way to destroy them both.[62] In fact, More's philosophy, with its echoes of paganism, is more dangerous than her own vitalism, which is not atheistic: "Neither do I think it Atheistical (as your Author deems) to maintain this opinion of self-motion, as long as I do not deny the Omnipotency of God; but I should rather think it Irreligious to make so many several Creatures as Immaterial Spirits, like so many severall Deities, to rule and govern Nature."[63]

Henry More not only believed in the existence of the Spirit of Nature; he also believed the world teemed with various immaterial beings. To Cavendish, therefore, his philosophy was anti-Christian: "But Madam, all these [stories about demons and genies], as I said, I take for fancies proceeding from the Religion of the Gentiles, not fit for Christians to embrace for any truth; for if we should, we might at last, by avoiding to be Atheists, become Pagans, and so leap out of the Frying-pan into the Fire, as turning from Divine Faith to Poetical Fancy; and if Ovid should revive again, he would, perhaps, be the chief head or pillar of the Church."[64] Although Cavendish did not deny the possibility of supernatural incorporeal spirits, she denied that such creatures could move inert matter: "No Angel, nor Devil, except God Impower him, would be able to move corporeal Matter, were it not self-moving, much less any Natural Spirit."[65] Thus, it is impossible, as More had argued, for the Devil to make contracts with witches because a supernatural being cannot empower a man to do material harm. Those spirits that may exist are like fairies, which are material, even though humans may not be able to perceive that they have "airy bodies, and be of humane shape, and have humane actions, as I have described in my Book of Poems."[66]

Cavendish, of course, had no problem with "Poetical Fancies, for that I am a great lover of them, my Poetical Works will witness; onely I think it not fit to bring Fancies into Religion."[67] Henry More made a category mistake when he integrated fancy and religion, a mistake Cavendish had always tried, if not always successfully, to avoid herself. Moreover, the breaching of categories implicit in Platonic philosophy, in which the supernatural is mixed with the natural, is a particular danger to women. Thinking that only their souls loved, women might fall into a more dangerous form of material adultery. Cavendish warns, "I am no Platonick; for this opinion is dangerous, especially for married Women, by reason the conversation of the Souls may be a great temptation, and a means to bring Platonick Lovers to a neerer acquaintance, not allowable by the Laws of Marriage, although by the sympathy of the Souls."[68]

Thus, the Neoplatonic philosophy of Henry More is not only a denigration of God and Nature; it is also a source of moral degeneration. Immaterialism leads to immorality. More had gotten it all wrong by trying to substitute an immaterial spirit of nature for the benevolent delegated rule of a female Nature. He had erred when he tried to understand nature through the lens of male preconceptions, substituting force for inherent power. He was mistaken in believing that Nature was a girl in need of protection. Nature is a woman, and she is the source of her own power.

## JOAN BAPTISTA VAN HELMONT

While Cavendish feared that Henry More was trying to subjugate Nature, she felt that Joan Baptista Van Helmont might drive Nature to seek her own death.

Writing about the Belgian alchemist—and his "obscure, intricate, and perplex" alchemical terms—she laments, "What with his Spirits, meer-beings, non-beings, and neutralbeings, he troubles Nature, and puzzles the brains of his Readers so, that, I think, if all men were of his opinion . . . Nature would desire God, she might be annihilated."[69] Of all the philosophers Cavendish considered in *Philosophical Letters*, she viewed Van Helmont as the weirdest, even more difficult to understand than Hobbes, Descartes, or More.[70] As in her discussion of More, Cavendish was particularly critical of the alchemist's immaterialism and the infusion of spiritual ideas into his philosophy, especially when they resulted in challenges to her own matter theory and concept of Nature. But in this text, even more than in her earlier disquisitions, Cavendish was aware of the potentially disastrous implications of the doctrine of immaterial substance on the perception and position of women.

Van Helmont was an iatrochemist, that is, a medical alchemist, and a follower of the sixteenth-century alchemist and doctor Paracelsus—of whose system one of the foremost historians of chemistry remarks, "Confusing? Of course. It was a puzzle to his sixteenth- and seventeenth-century followers as well as us today."[71] At the base of Van Helmont's revised Paracelsianism were principles such as the *semina*, which Walter Pagel defines as the "primordia from which all things created were to develop"; the *blas*, an "astral-cosmic force": and the *archei*, which are "the vital principles which 'conceive' generative ideas and images." In short, for Van Helmont, "the world consisted of enmattered psychoid impulses that were intrinsic, rather than superadded to matter. They accounted for its function and 'life.'"[72]

A modern reader can sympathize with Cavendish when she complains, "Who is able to conceive all those Chymaeras and Fancies of the Archeus, Ferment, various Ideas, Blas, Gas, and many more, which are neither something nor nothing in Nature, but betwixt both, except a man have the same Fancies, Visions, and Dreams, your Author had?"[73] Van Helmont's problem, according to Cavendish, is that his doctrine is "hermaphroditical," combining spirit and matter and maintaining something in-between. She does not seem to grant the alchemist the same flexibility in uniting reason and fancy that she allowed herself. His ideas are nothing more than "Hobgoblins," which "can do no great harm, except it be to trouble the brains of them, that love to maintain those opinions." As forces in nature, they are devoid of power because, just like Henry More's immaterial spirits, they cannot affect matter, for "nothing can do nothing."[74]

Cavendish argued that Van Helmont, who ascribed his knowledge of nature to dreams and illuminations, had created a system of neutral Being, which "is a strange Monster, and will produce monstrous effects."[75] The ideas of monsters in nature had significant import for the female philosopher. Liminal beings that were neither one thing nor another deeply disturbed her. She knew that she could be considered monstrous for doing what she did, and in her philosophy, she tried to include monsters in the natural taxonomy of nature. She also wanted to banish all occult and supernatural forces, although she admitted there were many internal aspects of natural things human beings did not understand.[76] Unlike Van Helmont, who believed the essence of nature could be understood by a kind of intuitive cognition of the imagination that allows union with objects, Cavendish believed that sense and reason were the only guides to understanding. Pagel describes Van Helmont's epistemology in the following way: "Objects, and phenomena of nature are, then, 'given' to the human intellect through an illumination which is comparable to prophetic vision; it is the true light that shows the way to the 'naked' being of things, by contrast with the deceitful advice derived from reason. Illumination prepares the mind for legitimate scientific analysis."[77] Cavendish could only dismiss this kind of belief: "Your Author is much for Intellect or Understanding, but I cannot imagine how Understanding can be without Reason."[78] Much of her critique of Van Helmont concerns her skepticism about his theories of knowledge of both natural and divine forces.

Cavendish believed that the spiritual forces, and angels and devils, that populate Van Helmont's version of the universe are only figments of his imagination, and unlike her own poetical fancies, they lack all claim to reality. Cavendish explains, "Truly, I must confess, I have had some fancies oftentimes of such pure and subtil substances, purer and subtiler then the Sky or Aetherial substance is, whereof I have spoken in my Poetical Works; but these substances, which I conceived within my fancy, were material, and had bodies, though never so small and subtil; for I was not able to conceive a substance abstracted from all Matter."[79] Even when she used her imagination, Cavendish thought materially. Clearly she could not discuss insubstantial being without her thoughts turning to fairies. This demonstrates the unity of her work, and of her vitalism, but also indicates how sensitive she was to charges that she had somehow incorporated immaterialism into her matter theory. And so she goes on to argue, "All Thoughts and Conceptions are made by the rational Matter, and so are those which Philosophers call Animal Spirits, but a material Fancy cannot produce immaterial effects, that is, Ideas of Incorporeal Spirits."[80]

One of the ideas that produce material effects, Van Helmont argued, is the impression of objects of desire on pregnant women, who are particularly prone to imagination. When pregnant women see terrible sights, they produce monsters, explains the alchemist. Cavendish replies, "But, as I said above, those births are caused by irregular motions [of matter], and are not frequent and ordinary; for if upon every strange sight, or cruel object, a Child-bearing woman should produce such effects, Monsters would be more frequent then they are. In short, Nature loves variety, and this is the cause of all the strange and unusual natural effects."[81]

Cavendish discusses pregnancy and birth several times in her critique of Van Helmont, particularly his argument that a pregnant woman's desire for cherries produces cherrylike marks on the infant.[82] Cavendish's critique of Van Helmont centers on a kind of epistemological feminism. She argues against the alchemist's doctrine that imagination produces ideas that in turn impress themselves on the vital spirit and affect body. Van Helmont had written that women were particularly prone to bewitching ideas, "for, says he, 'Women stamp Ideas on themselves, whereby they, no otherwise than Witches driven about with a malignant spirit of despair, are oftentimes governed or snatched away unto those things, which otherwise they would not, and do bewail unto us their own and involuntary madness.'" Against this, says Cavendish, "I cannot but take exception in the behalf of our Sex."[83] Cavendish was alert to the identification of women with witches, and at the heart of her attack on Van Helmont is the effort to drive nonmaterial forces from nature, because of the way they could be used not only to discredit her own philosophy but also to endanger women. Van Helmont claimed that "God doth love Women before Men," but his philosophy was insulting to women, and ultimately to Nature herself. She argues, "For though Nature is old, yet she is not a Witch, but a grave, wise, methodical Matron."[84]

The European witch craze was waning by the time Cavendish wrote *Philosophical Letters*, but belief in witches was still common. Joseph Glanvill, a member of the Royal Society and one of the few scholars who took Cavendish seriously, planned to send her a copy of one of his works defending the reality of witchcraft.[85] We have seen that in her discussion of a Travelling Man in *Natures Pictures*, she had introduced a witch without labeling her as evil and demonic. In this text, Cavendish is far more precise. When we see the unusual motions Nature produces in her love of variety, Cavendish argues,

by reason we cannot assign any Natural cause for them, are apt to ascribe their effects to the Devil; but that there should be any such devilish Witchcraft, which is made by a Covenant and Agreement with the Devil, by whose power Men do echaunt or bewitch other Creatures, I cannot readily believe. Certainly, I dare say, that many a good old honest woman hath been condemned innocently, and suffered death wrongfully, by the sentence of some foolish and cruel Judges, meerly upon the suspicion of Witchcraft, when as really there hath been so such thing.[86]

According to Cavendish, therefore, those poor old women who have been condemned to death are the victims of man's inability to understand that not everything in nature can be understood. There are indeed many occult forces in nature, but they are not supernatural; rather, they are natural motions we cannot comprehend. "I believe," says Cavendish, "natural Magick to be natural corporeal motions in natural bodies." They are the products of Nature: "Not that I say, Nature in her self is a Magicianess, but it may be called natural Magick or Witchcraft, meerly in respect to our Ignorance."[87] Women and Nature, therefore, can be saved from the dangerous gendered implications of Van Helmont's philosophy only by a proper understanding of Cavendish's materialism.

The argument Cavendish used against More about the nature of Nature is reiterated in her critique of Van Helmont: "Nature is the onely Mistress and cause of all."[88] She took particular exception to the alchemist's argument, which she quotes, that "nature first being a beautiful Virgin, was defiled by sin; not by her own, but by Mans sin, for whose use she was created." Such a claim is arrogant and presumptuous in making "man the chief over all Nature . . . when he is but a small finite part of Infinite Nature, and almost Nothing in comparison to it."[89] Nature cannot be ravished by man, who is subordinate to her.

Van Helmont's presumption developed into impiety when he claimed to be able to understand the genesis of human beings and their salvation by the birth of the Savior to a Virgin, thereby releasing man from the sin of the defilement of the original virgin, Nature. According to Van Helmont, as Cavendish paraphrases him, "Adam did ravish Eve, and defloured her by force, calling him the first infringer of modesty, and deflourer of a Virgin; and that therefore God let hair grow upon his chin, cheeks, and lips that he might be a Compere, Companion, and, like unto many four-footed Beasts, and might bear before him the signature of the same; and that, as he was lecherous after their manner, he might also shew a rough countenance by his hairs."[90]

Cavendish knew that there was nothing in scripture about this particular consequence of the Fall and that Van Helmont was presuming to make a "romance" out of Genesis. And of course Cavendish's philosophy had a much more favorable understanding of animal nature, which is not bestial but rational in its own way. Moreover, Cavendish argues, the comeliness of "Eve's Daughters prove rather the contrary, viz. that their Grandmother did freely consent to their Grandfather." Likewise, when Van Helmont argues that all the monsters in nature, including

"Faunes, Satyrs, Sylphs, Gnomes, Nymphs, Driades, Nayades, Nereides, etc.," are the products of the original copulation of men with the monstrous "daughters of men" who were created by Satan, in "the womb of a Junior Witch or Sorceress," he was reading into scripture and demeaning God's power.[91] God's design could not be perverted by either female monsters, witches, or the Devil, and monsters are simply Nature "taking delight in variety."[92]

Cavendish rejected the association of the female with sin and sorcery, even if Adam and his fellow men were the perpetrators of rape and bestiality. In early modern law, in most cases, rapists of unmarried women were sentenced leniently, and if the woman conceived after the assault, it was considered evidence of her complicity.[93] Cavendish knew that most people blamed Eve for original sin, "as others do lay all the fault upon the Woman, that she did seduce the Man." Even if Van Helmont exonerated Eve of the crime, she would share the guilt for most of his readers. While the alchemist "shews a great affection for the Female Sex" and believes it particularly suited to a devotion to God, he is wrong to think that more women than men are born, or that more men than women are punished with death.[94] Men may die in wars and duels, but a like number of women die in childbirth. Women and men are equal in their mortality and presumably equal in their creation.

The height of Van Helmont's presumptuous retelling of scripture, according to Cavendish, comes in his gynecological discussion of "the Conception of the Blessed Virgin." The alchemist

> Doth not stick to describe exactly, not onely how the blessed Virgin conceiv'd in the womb, but first in the heart, or the sheath of the heart; and then how the conception removed from the heart into the womb, and in what manner it was performed. Certainly, Madam, I am amazed, when I see men so conceited with their own perfections and abilities, (I may rather say, with their imperfections and weaknesses) as to make themselves God's privy Councellors, and his Companions, and partakers of all the sacred Mysteries, Designs, and hidden secrets of the Incomprehensible and Infinite God. O the vain Presumption, Pride, and Ambition of wretched Men![95]

Cavendish did not often sound like a Calvinist preacher, but Van Helmont's exploration of the female role in sin and salvation, and his attempt to penetrate into divine mysteries, was too much for her. He and those like him, by "all their strict and narrow pryings into the secrets of God, are rather unprofitable, vain and impious, then that they should benefit either themselves, or their neighbor."[96]

Cavendish believed that Van Helmont and his kind not only insulted God but also demeaned Nature when they tried to force Nature's secrets out of her with experiments or "Art." She argues, "Nor can Art steal from Nature; she may trouble Nature, or rather make variety in Nature, but not take any thing from her, for Art is the insnarled motions of Nature: But your Author, being a Chymist, is much for the Art of Fire, although it is impossible for Art to work as Nature doth; for Art makes of natural Creatures Artificial Monsters, and doth oftner obscure and disturb Natures ordinary actions then prove any Truth in Nature."[97]

Cavendish's criticism of Van Helmont's alchemy is the opening round in her attack on experimental philosophy, to which she will devote an entire book, the *Observations upon Experimental Philosophy*, published two years after *Philosophical Letters*. The theme of that work, and *New Blazing World*, is foreshadowed here. Experimental philosophy produces artificial monsters and is essentially useless. When "chymists" or experimenters try to imitate Nature and expose her secrets, they are presumptuous and foolhardy. Although Nature loves all diversity in nature, the manufactures of either chemical manipulation or experimental art are secondary to Nature's own productions. Art is "Nature's Mimick or Fool"; its adherents are therefore, by implication, the jesters who play in the Royal Society, which Cavendish held up to laughter in her visit.[98]

But natural philosophy is not a laughing matter. In the rest of *Philosophical Letters*, Cavendish pursues her theme that Nature herself is endangered by the ideas and actions of her investigators. Mathematicians, she states, "endeavour to inchant Nature with Circles, and bind her with lines so hard, as if she were so mad, that she would do some mischief, when left at liberty." In other words, these men are witches and torturers. Geometers, with their weighing and measuring, "do press and squeeze her so hard and close, as they almost stifle her." Natural philosophers try to "stuff her with dull, dead, senceless *minima's*, like as a sack with meal, or sand, by which they raise such a dust as quite blinds Nature and Natural Reason." But of all the tormentors of Nature, "Chymists torture Nature worst of all, for they extract and distil her beyond substance, nay, into no substance, if they could."[99] It seems that, for Cavendish, all philosophy was an assault on at least one woman, who was tortured, pressed, blinded, and executed by its male proponents. The philosophers may have not been consciously sexist in their attitude toward nature, but Cavendish realized the sexual implications for both women and nature of the new philosophy.

All the philosophers Cavendish considered in *Philosophical Letters* were guilty of one supreme crime: in their arrogance they thought that they understood Nature and could manipulate her. Hobbes thought that nature was determined by the force and impacts of its particles. Descartes adopted this mechanistic model and further diminished material nature by depriving it of life and motion. His dualism expanded the power of incorporeal substance at the expense of all natural being. Henry More's immaterialism was even worse: his Spirit of Nature was an attack on the independence and power of Nature; in a sense he tried to infantilize Nature. His immaterial spirits were pagan and an affront to the majesty of both God and Nature. But worst of all was Van Helmont, whose impious assault on nature, by invoking substantial but incorporeal powers, ultimately attacked women by linking the female, the monstrous, and the witch.

When Cavendish ventured into male territory in *Philosophical Letters*, she found an undiscovered country not at all congenial to its visitor. In her next work, she envisioned a voyage to a new world, where the arrogance and presumption of her male fellow natural philosophers are curtailed and put to the use of her own projected self. In *Observations upon Experimental Philosophy*, Cavendish presented a detailed defense of Nature within the context of a critique of Robert Hooke's *Micrographia*. She attacked the Royal Society for ripping nature apart and assuming they had the right to do so because of their sex. In *Blazing World*, which parodies the pretensions of experimental philosophers, Cavendish created a fantasy where an Empress is worshiped by beast-men scientists and served by immaterial spirits. At least in fantasy, her position as the most powerful woman in the world is secured, and the governance of a female being is made concrete.

# chapter seven

# Cavendish against
# the Experimenters

*xperimental philosophy presented Margaret Cavendish with a problem. In
*Philosophicall Fancies*, she had condemned alchemy but used some of its terminology to explicate her emerging natural philosophy. In *Philosophical Letters*, she
attacked the Paracelsian notions of Van Helmont and equated the manipulation
of nature with torture. But Cavendish also enjoyed a long correspondence from
1657 until 1671 with the Dutch diplomat and man of letters Constantijn Huygens
(1596–1687), the father of the future scientist Christiaan Huygens (1629–95). They
discussed experimental topics, including the phenomenom of Rupert's drops,
molten glass drops that fragmented in an unusual way depending on whether
force was applied to their tails or bodies.[1] In the 1630s and 1640s, Newcastle was
also interested in alchemy and glass and even wrote recipes for chemical medi-
cines.[2] But by 1664, in *Sociable Letters*, Cavendish lamented that an ordinary
"glass-ring" caused most people to "Wonder how it came to be Invented, and Admire the Inventor for a Person of an Ingenious Brain," but at the same time, she
reacted to "any One person that had the Ingenuity to Invent Arts, or Find out
New Sciences . . . or the Deep Conceptions of Philosophy" with "Yawning,
Humming, Hauking and Spitting."[3]

Cavendish increasingly saw herself and her work in opposition to experimental philosophy. The experimenters were stealing her audience and indeed turning all their admirers into "mongrels . . . half Men and half Beasts, or Dull, Ignorant Persons, as half Men, and half Stones or Blocks."[4] In *Observations*, she shows
little sympathy for those who tamper with Nature, and in *Blazing World*, she uses
the image of hybrid beast-men to ridicule both the experimenters and their
claims to knowledge of the material world. Her audience will be the future, she
explains to her readers, when instead of being "slighted now and buried in silence, she may perhaps rise more gloriously hereafter; for her Ground being

Sense and Reason, She may meet with an age where she will be more regarded, then she is in this."[5]

Cavendish did not expect the practitioners of experimental philosophy, or as she put it, "mode-philosophy," to treat her ideas with civility. To some extent, they share the "Brutish Nature" of their audience. The polemical discourse that characterized *Philosophical Letters* continued in an all-out onslaught on the social and intellectual pretensions of the experimental philosophers in *Observations upon Experimental Philosophy* (1666)—a book inspired by the publication of Robert Hooke's *Micrographia* (1665), the Royal Society's signature text. Cavendish portrayed her attack as a duel, the prize being the recognition of her status and achievements by "the impartial world." "I am," she writes, "as ambitious of finding the truth of Nature, as an honourable dueller is of gaining fame and repute." It is an act of courage, she claims, to challenge both speculative philosophers and "our Modern Experimental and Dioptrical Writers."[6] But her contemporary opponents are not honorable; in fact, they are arrogant and lacking in civility. "They will perhaps think my self an inconsiderable opposite, because I am not of their Sex, and therefore strive to hit my Opinions with a side stroke, rather covertly, then openly and directly." But "the impartial World, I hope, will grant me so much Justice as to consider my honesty, and their fallacy, and pass such a judgment as will declare them to be Patrons, not onely to Truth, but also to Justice and Equity."[7]

While the speculative philosophers Cavendish attacked in her earlier works might have been impious or even heretical, she arraigns the modern philosophers as guilty of what seventeenth-century society would consider an even more deplorable trait: they are not honorable. Honor itself meant to be honored by others, to be recognized as a gentleman, to possess virtue, honesty, and civility.[8] Honor was sought by all members of society trying to establish their place in the hierarchy of early modern Europe. It was equally a currency among natural philosophers seeking renown within their own constellation of peers, the larger reading public, and polite society.

Traditionally, honor was established through patronage, the informal system of interlinking and mutually beneficial relationships that held together the social structure of the time. Unless a client's honor was acknowledged by a superior, who benefited from the client's support and actions, his chances of success were minimal. A thinker's intellectual successes reflected his social acceptability. Ultimately, when Cavendish challenged the honor and civility of the experimental philosophers, she was contesting both their social and intellectual claims at a time when these investigators were trying to establish the legitimacy of their profession, their method, and their ideas.

Patronage was a uniquely potent weapon for a woman. Although women were blocked from participation in the formal institutions of government and education, patronage gave them access to power and learning. Descartes had used the support of Princess Elisabeth of Bohemia and Queen Christina of Sweden, Galileo sought the patronage of Duchess Christina of Tuscany, and Henry More depended on the friendship of Lady Anne Conway. To some extent, these relationships empowered both patrons and clients, both the masters and their noble female students. In *Observations upon Experimental Philosophy* and *Blazing World*, Cavendish fashioned herself as both patron and client, claiming the status of worthy opponent in a duel of philosophers.

But in early modern England, honor also meant something different for a woman than for a man. Cavendish was well aware of the sexual connotations of honor. In her autobiography, describing her experience at the court of Henrietta Maria, she writes, "Though I might have learnt more Wit, and advanced my Understanding by living in a Court, yet being dull, fearfull, and bashfull, I neither heeded what was said or practic'd, but just what belong'd to my loyal duty, and my own honest reputation; and indeed I was so afraid to dishonour my Friends and Family by my indiscreet actions, that I rather chose to be accounted a Fool, than to be thought rude or wanton."[9]

In defending her own honor, and in disparaging the potentially dishonorable actions of the experimenters, Cavendish was also defending her reputation as an honest woman. She had already problematized her honor by displaying her works in public. Mary Evelyn commented, about meeting Cavendish, that her manner was "airy, empty, whimsical, and rambling as her books, aiming at science, difficulties, high notions, [but] terminating in nonsense, oaths, and obscenity."[10] Evelyn had met Cavendish during her trip to London when she put herself on display before the Royal Society, where she overcame her fear of being perceived as a fool by making her hosts appear foolish. She protected her honor, both sexual and philosophic, on this occasion by fashioning herself as a royal woman on a procession. This metamorphosis from maiden to monarch paralleled the fictional transformation of a young lady, kidnapped by force by a lower-class merchant, into the Empress of the New Blazing World.

Cavendish invoked several strategies to undermine the pretensions of the members of the Royal Society and substantiate her own claim to be a natural philosopher—and a woman—worthy of honor. *Observations*, which contains little new in terms of her natural philosophy, was intended as part of a patronage strategy to gain favor for her work while demonstrating the fallacies of the new science. Employing the protocols of patronage, she attempted to undermine the authority of the Royal Society. She presented what she thought was a more probable picture of nature than the deluding instruments of experimental science could offer. She portrayed the experimenters themselves as boys playing with toys or, more worryingly, as potential revolutionaries, who by their arrogance challenged the authority of the schools and the state. She implied, in fact, that the members of the Royal Society were like women, mixing and brewing to produce artificial and hermaphroditical effects. Finally, having exhausted the polemical potential of straight criticism, Cavendish in *Blazing World* turned to fantasy to confirm her position as the patron of knowledge who rules her imagined world's monstrous metamorphosed experimental philosophers. In this work, the new scientists' lack of civility is converted into a lack of humanity: experimenters become bear-men and chemists become ape-men. Cavendish reveals a new world that can be discovered only through the imagination, not the experimental philosophy of the Royal Society.

## HONOR, PATRONAGE, AND POLITICS

From the beginning of her career, Cavendish fashioned herself as a client seeking the support of a patron for her literary and scientific endeavors. Her earlier works are peppered with letters, epistles, and prefaces urging her readers to approve her works and acknowledge her abilities and claiming a superior authority and understanding compared with modern natural philosophers. In her *Sociable Letters* and *Philosophical Letters,* she even created her own reader/alter ego to whom she submitted the protection of her works. "Thus, Madam," she wrote, "although I am destitute of the help of Arts, yet being supported by your Favour and wise Directions, I shall not fear any smiles of scorn, or words of reproach; for I am confident you will defend me against all the mischievous and poisonous Teeth of malitious detractors."[11]

But such self-generated and self-regarding support was perhaps not the most viable way of entering the philosophic community. As in the 1655 edition of *Opinions,* Cavendish's most tangible patronage strategy was an appeal to the universities of Europe, particularly the University of Cambridge.[12] By privileging the authority and preeminence of the university, Cavendish hoped to subvert the efforts to keep her out of the community of natural philosophers. In particular, Cavendish emphasized the "civility" of the university. She wrote, in response to a thank-you letter from the chancellor of the university,

> You might, if not with scorn, with silence have passed by, when one of our Sex, and what is more, one that never was versed in the sublime Arts and Sciences of literature, took upon her to write, not onely of Philosophy, the highest of all humane Learning, but to offer it to so famous and celebrated a University as yours; but your Goodness and Civility being as great as your Learning, would rather conceal, then discover or laugh at those weaknesses and imperfections which you know our Sex is liable to; nay, so far you were from this, that by your civil respects, and undeserved commendations, you were pleased to cherish rather, then quite to suppress or extinguish my weak endeavours.[13]

The members of the university, therefore, unlike the modern philosophers, were gentlemen who knew how to treat a lady, even one with pretensions to knowledge. Moreover, the university as the upholder of civility or polite society had the power, according to Cavendish, to admit her within that civilization of the learned. Without its legitimizing power, Cavendish recognized, she would not be courted or included within sage society—here meaning, quite literally, the society of the sage. Therefore, her only path to recognition and honor would be through its patronage. As a client in need of a protector, Cavendish elevated the status of the patrons she sought, calling them the stars governing the whole world of the learned. Such recognition, in turn, gave that patron—the university—the power to elevate her own status and honor. This mutual benefit of patronage was recognized by both parties. The vice chancellor of Cambridge wrote to Cavendish in 1668 that the present age "will not judge us to be no-bodies whom such an accomplish'd Princess hath not refused to make not only the Perusers, but even the Moderators, and Judges of her works."[14]

In the contest for honor and legitimacy in the 1660s, the Royal Society was trying to make the university professors "no-bodies," not worthy of the social preeminence the society claimed. At the same time, the experimenters tried to fashion a new identity for themselves. As we saw in Chapter 1, experimental philosophers had an image problem. On the one hand, they could be linked with the entertainers who displayed wonders at country fairs. On the other, they could be associated with university professors—whose social status, ironically, was problematic. As Steven Shapin has argued, "The characters of the pedant and the gentleman were set in radical opposition." Thus, the Royal Society was desperately trying to articulate a new ideal of the gentleman-scholar.[15]

In seeking status, the society itself had sought the patronage of the king and the Earl of Clarendon, among others. Many of the gentlemen virtuosi belonging to the society had been recruited for the social legitimacy they would bring.[16] And there was the real advantage that one of the foremost practicing scientists of the society was Robert Boyle, brother of the Earl of Cork and an uncontestable gentleman.

Cavendish, in *Observations*, refers to several experiments and claims Boyle had made in *Experiments and Considerations Touching Colours*, although he is not specifically named. She critiques Boyle on the tempering of steel, the nature of light and dark, the nature of refraction, the properties of snow, and a blind man's ability to feel color.[17] In the chapter on colors in *Observations*, she dismisses the theory of color and light Boyle had presented in *Experiments*. The only time she mentions him explicitly is in *Philosophical Letters* (1664), published in the same year as *Experiments*. She refers to "the Learned and Ingenious Writer B," some of whose works she has read: "I have observed he is very civil, eloquent, and rational Writer; the truth is, his style is a Gentleman's style. And in particular, concerning his experiments, I must needs say this, that, in my judgment, he hath expressed himself to be a very industrious and ingenious person; for he doth neither puzle Nature, nor darken truth with hard words . . . besides, his experiments are proved by his own actions." But she suggests that these experiments might be more useful if Boyle would study the alterations of parts caused by motion rather than just the alterations themselves. Then, certainly, "he might arrive to a vast knowledg by the means of his experiments; for certainly experiments are very beneficial to man."[18] In this case, Boyle's social status trumps his methodology, although in other places of *Philosophical Letters* Cavendish foreshadows the harsh critique of experimental philosophy that appears two years later in *Observations* and *Blazing World*.

Cavendish probably had also read Boyle's *The Sceptical Chymist* (1661). In it, she would have seen his claim "that to keep a due decorum in the discourses it was fit that in a book written by a gentleman, and wherein only gentlemen are introduced as speakers, the language should be more smooth and the expressions more civil than is usual in the more scholastic way of writing."[19]

Tracing the fine rhetorical line Boyle draws between a gentleman and a university professor, one can sense how problematic the claim to civility—to being a gentleman—might be for experimenters in the late seventeenth century. Obviously, the claim of possessing civility was very much on the minds of the experimentalists and their opponents. While Boyle might be civil, according to Cavendish, when most modern philosophers write about the ancients, "they reward them with scorn, and rail at them. . . . To which ungrateful and unconscionable act, I can no ways give my consent, but admire and honour both the ancient, and all those that are real Inventors of noble and profitable Arts and Sciences, before all those who are but botchers and brokers."[20]

Cavendish condemned the Royal Society for its high-handed and presumptuous behavior. This charge was particularly challenging because the society prided itself on its civility and lack of dogmatism, at least during meetings. Such politeness, the members believed, would protect it from the potential chaos engendered by too much passion or enthusiasm, excesses that had corrupted both the universities and the state in the members' very recent memories.[21] Cavendish perceived how the society was trying to establish its bona fides by claiming to be a new social space different from that of the potentially disruptive universities. She also understood that it was trying to seize from the schools the prerogative of conferring intellectual validity. She charged, "Many will not admit of rational arguments, but the bare authority of an Experimental Philosopher is sufficient to them to decide all Controversies, & to pronounce the Truth without any appeal to Reason; as if they onely had the Infallible Truth of Nature, and ingrossed all knowledg to themselves."[22]

The Royal Society developed a new way to establish its honor, and its authority, a strategy sidestepping the traditional patronage system. By validating its own work and privileging it and the instruments involved as the most useful accomplishments of the modern age—comparable to the discovery of the New World—the society did not have to look outside itself to confirm its authority. But in Cavendish's eyes, by trying to make themselves the ultimate arbiters of natural philosophy, society members were not only irreverent to authority but also in rebellion against it. They were introducing civil war into the society of the learned, a civil war directed against the institutions of English society as well as against the learning of the ancients. In the preface to *Observations*, Cavendish exclaimed,

> But such Writers are like those unconscionable men in Civil Wars, which endeavour to pull down the hereditary Mansions of Noble-men and Gentlemen, to build a Cottage of their own; for so do they pull down the learning of Ancient Authors, to render themselves famous in composing their Books of their own. But though this Age does ruine Palaces, to make Cottages; Churches, to make Conventicles; and Universities to make Colledges; and endeavour not onely to wound, but to kill and bury the Fame of such meritorious Persons as the Ancient were, yet, I, hope God of his mercy will preserve State, Church, and Schools, from ruine and destruction.[23]

The charge that natural philosophers were rebels was a grave one in the period of the Restoration, when Charles II had just regained his throne from the rebels and regicides who had governed England during the civil war period. Newcastle, at the moment Cavendish was publishing this text, was trying to repair the devastation rebels had caused to his estates of Welbeck Abbey and Bolsover Castle. In light of her experience, even her own doubts about the validity of Aristotelianism did not justify attacking the authority of the ancients and those who taught ancient philosophies. In the dedication to the University of Cambridge in *Philosophical Letters*, Cavendish had in fact expressed the hope that the university's "several studies may be, like several Magistrates, united for the good and benefit of the whole Commonwealth, nay, the whole World."[24] In this case, the implication is that the experimental philosophers are not only disrupting the community of the learned but are a political danger in the Commonwealth and must be contained by the magisterial powers of the university. Once again, we are brought back to the problem of power and who possesses it. Cavendish seeks a judiciary of the learned, with the university as the judge possessing the power and authority to legitimize her work and acknowledge her status as a natural philosopher and, at the same time, to control and delegitimize the power and position of the experimental philosophers.

Cavendish, in the statement above, equates the pulling down of palaces to make cottages with the ruin of the universities to make private colleges. This statement may be an express reference to Gresham College, where the Royal Society had been meeting, or to the several proposals to create research colleges circulating in the early 1660s. In 1667, a year after *Observations upon Experimental Philosophy* was written, the Royal Society itself undertook a project to fund a private college with subscriptions from the general public. Later in 1667, after her famous visit to the society, Cavendish was asked to subscribe to a building fund for the new college. Not surprisingly (at least to us), she declined the honor.[25]

Of course, Cavendish also knew that while she might gain the patronage of the University of Cambridge, she would no more be able to become a university student than to become a member of the Royal Society. Her sex confined her to outsider status in all the societies of learning. There could be no parity for her either socially or intellectually; no man would engage her in a duel of swords or words. The result of her exclusion, she believed, was the lack of clarity of her philosophic works: "But as for Learning, that I am not versed in it, no body, I hope, will blame me for it, since it is sufficiently known, that our Sex is not bred up to it, as being not suffer'd to be instructed in Schools and Universities."[26]

Nevertheless, making a virtue out of necessity, Cavendish claimed her "natural wit" would compensate for her lack of knowledge of the "artificial" terms of the philosophers. The benefit would be not only philosophic but social: "Although I do understand some of their hard expressions now, yet I shun them as much in my writings as is possible for me to do, and all this, that they may be the better understood by all, learned as well as unlearned."[27] Cavendish's audience, as well as her ultimate patron, is the whole world, which she begs to treat her ideas with "justice and equity." In return, she proffers, "I will not deceive the World, nor trouble my Conscience by being a Mountebanck in learning, but rather prove naturally wise then artificially foolish."[28] By equating the jargon-using philosophers with mountebanks, Cavendish explicitly reveals the strategy she will use in *Observations* and *Blazing World*.[29] Rather than being worthy of honor and the patronage of the universities or the world, the experimenters are fools, worthy not of admiration but contempt.

## EXPERIMENTERS AS FOOLS, OR WORSE

If the experimental philosophers are not gentlemen, then what are they? Cavendish has a suggestion in *Observations*: "But as Boys that play with watry Bubbles ["Glass-tubes"], or fling Dust ["Atomes"] into each others Eyes, or make a Hobbyhorse ["Exterior Figures"] of Snow, [they] are worthy of reproof rather then praise; for wasting their time with useless sports; so those that addict themselves to unprofitable Arts, spend more time then they reap benefit thereby."[30] Not only do these boys (and remember that Cavendish saw herself and Nature as mature women) infantilize themselves by pursuing the "superficial wonders" of nature, but they fail to do anything useful. Cavendish admits that if they could achieve anything of benefit to humankind, such as improve agriculture or architecture, or find ways "for the decrease of nice distinctions and sophistical disputes in Churches, Schools and Courts of Judicature, it would not onely be worth their labour, but of as much praise as could be given to them."[31] But, instead, experimenters make "pretty toys to imploy idle time."[32] Instead of seriously contemplating nature, they practice "deluding arts" that only obfuscate human knowledge.

In fact, she concedes, the experimenters do produce a kind of second order of natural effects. To the extent that they manipulate the natural world, their "artificial effects" are in some sense part of nature and can even be useful in the right circumstances. However, an artificial effect can be neither as good nor as lasting as a natural one. Indeed, according to Cavendish, "They are but Natures bastards or changelings, if I may so call them; and though Nature takes so much delight in variety, that she is pleased with them, yet they are not to be compared to her wise and fundamental actions; for Nature, being a wise and provident Lady, governs her parts very wisely, methodically and orderly."[33]

Thus, the boyish experimenters are concerned only with Nature's by-products, her illegitimate and inhuman productions. They denigrate the absolute glory of Nature by concentrating on what Cavendish calls "her works of delight, pleasure and pastime: Wherefore those that imploy their time in Artificial Experiments, consider onely Natures sporting or playing actions."[34] Rather than addressing Nature's most fundamental creations, the experimenters make Nature into a kind of dilettante or virtuoso like themselves. In contrast, those who (like Cavendish) "view her wise Government, in ordering all her parts, and consider her changes, alterations and tempers in particulars, and their causes, spend their time more thoughtfully and profitably; and truly to what purpose should a man beat his brains, and weary his body with labours about wherein he shall lose more time, then gain knowledg?"[35]

The experimenters, according to Cavendish, are committing a kind of lèse-majesté when they claim to be able to understand and imitate Nature's actions. They attempt to usurp the power of Nature, without realizing that only Nature herself controls her own parts, including the human parts. Cavendish may have had Robert Boyle specifically in mind when it came to the question of human presumption. In his *The Usefulness of Natural Philosophy* (1661), Boyle had argued that "the Study of Physiology is not only Delightful, as it teaches us to Know Nature, but also as it teaches us in many Cases to Master and Command her." Indeed, argues Boyle, the "Dominion, that Physiologie gives the prosperous Studier of it . . . is a Power that becomes Man as Man." The enjoyment of this control of nature "may appear in the Delight Children take to do many things . . . that seem to proceed from Innate Propensity to please themselves in imitating or changing the Productions of Nature."[36]

Cavendish can reply to this claim only by reversing it. Men (and boys) are Nature's unwitting agents, who even in their conspiracies against her are her pawns. Man's power disappears before Nature: "But neither can natural causes nor effects be over-powred by man so, as if man was a degree above Nature, but they must be as Nature is pleased to order them; for Man is but a small part, and his powers are but particular actions of Nature, and therefore he cannot have a supreme and absolute power."[37]

It is important to note the gendered content of Cavendish's conclusion in the above statement. In a much earlier pronouncement in *The Worlds Olio*, Cavendish asked about the position of women: "What ever did we do," she asks, "but like Apes, by Imitation?" The oppression of men "hath so dejected our spirits, as we are become so stupid, that Beasts are but a Degree below us, and Men use us but a Degree above Beasts."[38] Arrogant experimental philosophers are just another aspect of the masculine subjugators of women, who attempt to dominate Nature as men dominate women. In this case, they do not succeed. Cavendish states, "But I perceive, Man has a great spleen against self-moving corporeal Nature, although himself is part of her, and the reason is his Ambition; for he would fain be supreme and above all other Creatures, as more towards a divine Nature: he would be a God, if arguments could make him such, at least God-like, as is evident by his fall, which came meerly from an ambitious mind of being like God."[39] In fact, experimentation "strives to imitate Nature, yet it is so far from producing natural figures, that at best, it rather produces Monsters instead of natural effects."[40]

In this text, it turns out that the art (or experimentation) of the experimental philosophers "is like an emulating Ape." In trying to imitate Nature, experimenters create a science that apes Nature's actions. Paradoxically, the metaphor Cavendish had used to describe women now becomes more appropriate for the male experimenters. These feminized men generate "a praeter-natural or irregular production": "not praeternatural in respect to general Nature, but in respect to the proper and particular nature of the figure. And in this regard I call Artificial effects Hermaphroditical, that is, partly Natural, and partly Artificial; Natural, because Art cannot produce any thing without natural matter, nor without the assistance of natural motions, but artificial, because it works not after the way of natural productions."[41] In Nature, human beings are produced in a regular fashion, "by the copulation of two persons of each Sex," but in the art of the experimental philosophers, their irregular experiments produce monsters. Indeed, even if experimenters could manufacture a flying man, "we should rather account that Man Monstrous that could flie, as having some motion not natural and proper to his figure and shape." Every "Creature is perfect in its kind, that has all the motions which are naturally requisite to the figure of such a kind." Nature has created all of its parts perfectly, with due order and harmony, "but Man is apt to run into extreams, and spoils Nature with doting too much upon Art."[42]

The ultimate problem of experimental philosophy is that it tries to go where no man can go. In attempting to reveal the interior of natural things with their microscopes and telescopes, experimenters only distort Nature and her creatures. Indeed, Cavendish concedes, while an instrument can present an accurate picture of an object, "yet that natural figure may be presented in as monstrous a shape, as it may appear mis-shapen rather then natural: For example; a Lowse by the help of a Magnifying-glass, appears like a Lobster. . . . The truth is, the more the figure by Art is magnified, the more it appears mis-shapen from the natural, in so much as each joynt will appear as a diseased, swell'd and tumid body, ready and ripe for incision."[43] It is clear from this unusually graphic comment that Cavendish has been studying the illustrations in Hooke's *Micrographia*, which include a flea and a louse. But although she finds them to be "wonders" worthy of "admiration," they are also corrupting and disgusting. And they are threatening, not only for the animals portrayed but also for women. Cavendish comments, "Had a young beautiful Lady such a face as the Microscope expresses, she would not onely have no lovers, but be rather a Monster of Art, then a picture of Nature, and have an aversion, or at least a dislike to her own exterior figure and shape; and perchance if a Lowse or Flea, or such like insect, should look through a Microscope, it would be as much affrighted with its own exterior figure, as a young beautiful Lady when she appears ill-favoured by Art."[44]

In adopting the perspective of the objects of microscopic inspection, Cavendish reveals how dangerous experimental philosophy can be. The experimenters are just waiting to rip apart the objects they view; possibly the beautiful young lady will be the next object of their penetration. Lice, fleas, and women are all dehumanized by observation, to the extent that even in their own eyes they would be monsters. The microscope creates hybrid figures out of nature and art, which transgress the borders between the human and the bestial; it manufactures monsters of perception.

Thus, in trusting to sense rather than reason, "Experimental Philosophy has but a brittle, inconstant and uncertain ground, and these artificial Instruments, as Microscopes, Telescopes, and the like, which are now so highly applauded, who knows, but may within a short time . . . be found deluders rather then true Informers."[45] Cavendish spends more than a hundred pages in *Observations* discussing perception and its inability to comprehend the "interior, proper, and innate actions" of a creature.[46] She argues, the dissection of "a living Creature can no more inform one of the natural motions of that figure, then one can by the observing of an egg, be it never so exact, perceive the corporeal figurative motions that produce or make the figure of a Chicken." In fact, "it is as improbable for humane sight to perceive the interior corporeal figurative motions of the parts of an animal body by Anatomy, as it is for a Micrographer to know the interior parts of a figure by viewing the exterior."[47] So much for trying to peel away the skins of objects—ladies and lice, not to mention chickens, are immune from the penetrating gaze of the experimental philosophers.

The inability of experimenters to view natural objects reveals the emptiness of their enterprise. The new philosophy with its exaggerated picture of lice cannot even help a beggar, even if he could see its image in a microscope, "for it doth neither instruct him how to avoid breeding them, or how to catch them, or to hinder them from biting."[48] Experimental art might become a useful servant of nature, Cavendish concedes, if it relied more on reason, but "in this age she is become rather vain then profitable, striving to act beyond her power." In fact, in an ironic counterpoint, Cavendish admits she too is guilty of trying "to write beyond my experience, for which, 'tis probable Artists will condemn me." In asking the experimenters' pardon, Cavendish again underscores the limits of experimental philosophy: "[I] pray them to consider the Nature of our sex, which makes us, for the most part, obstinate and wilful in our opinions, and most commonly impertinently foolish: And if the Art of Micrography can but find out the figurative corporeal motions that make or cause us to be thus, it will be an Art of great fame, for by that Artists may come to discover more hidden causes and effects; but yet I doubt they will hardly find out the interior nature of our sex by the exterior form of their faces or countenances."[49]

The mysterious female eludes the understanding of the investigating man; he can never understand why she is what she is. Nevertheless, there is a role for women to play in experimental philosophy. Women, like the "good huswife Nature," are good at making artificial things like "Sweetmeats, Possets, several sorts of Pyes, Puddings, and the like." Hence, "they would prove good Experimental Philosophers, and inform the world how to make artificial Snow by their Creams or Possetts beaten into froth, and Ice by their clear, candied or crushed quiddinies . . . and many other the like figures which resemble Beasts, Birds, Vegetables, Minerals, etc." It is only natural that women should take on the task of experimenting from men, "for the Woman was given to Man not onely to delight, but to help and assist him." Their labor would release men to study causes and thereby help the Commonwealth, "and then would Men have reason to imploy their time in more profitable studies, then in useless Experiments."[50]

The sarcasm that will become the basic tone of Blazing World can be heard in this suggestion of men giving up their experiments to women. By playing with the categories of male/female, Cavendish manages to ridicule men while condemning most women. Women do not even eat the puddings they make; it is just a way "to imploy their idle time." Their creations are like the toys experimenters make in their "idle time." But at least this kind of activity is natural for women. Men, in assuming female kinds of activity, are essentially becoming unnatural. They are trying to produce effects that can only be hermaphroditical because "no

Creature by any Art whatsoever, is able to produce a new form, no more then he can make an new atome of new matter, by reason the power lies in Nature, and the God of Nature, not in any of her Creatures." There is nothing new under the sun, for "nothing is created anew, which never was in Nature before."[51]

Not only are experimenters wrong in their methodology, but they are mistaken in their ontology, also. In another reversal of roles, many new scientists, including Robert Hooke and Robert Boyle, had adopted the corpuscularism Cavendish abandoned after her first book. "But the opinion of Atomes," she argues, "is fitter for a Poetical fancy, then for serious Philosophy." If atoms existed and they were autonomous and individuated bodies, then "Nature would be like a Beggars coat full of Lice." Apparently haunted by the monstrous louse of *Micrographia*, Cavendish saw the material principles of the experimental philosophy as an infestation on the body of Nature. Worse, the atoms would disrupt the order of Nature, producing more "a confusion, then a conformity in Nature, because all Atomes, being absolute, they would all be Governours, but none would be governed."[52]

But, Cavendish contends, such a world could not come into being, even with the artificial tools of the experimental philosophy. "Indeed Art can no more force certain Atomes or Particles to meet and join to the making of such a figure as Art would have, then it can make by a bare command Insensible Atomes to join into a Uniform World."[53] Experimenters cannot make worlds or at least natural worlds, Cavendish argues, for

> if they should make a new world by the Architecture of Art, it would be a very monstrous one: But I am sure art will never do it; for the world is still as it was, and new discoveries by Arts, or the deaths and births of Creatures will not make a new world, nor destroy the old, no more then the dissolving and composing of several parts will make new Matter; for although Nature delights in variety, yet she is constant in her ground-works; and it is a great error in man to study more the exterior faces and countenances of things, then their interior natural figurative motions, which error must undoubtedly cause great mistakes, in so much as mans rules will be false, compared to the true Principles of Nature.[54]

The true principles of nature, of course, are the ones that Cavendish has discovered through her own reason and sense. If she could, Cavendish argues, she would "set up a sect or School for my self" to teach her doctrines, but "being a woman, do fear they would cast me out of their Schools." The presumptuous moderns would, if they could, even turn "the Muses, Graces and Sciences . . . from Females into Males; so great is grown the self-conceit of the Masculine, and the disregard of the Female sex."[55] Cavendish's only option is to create a world where she can create and rule scientific societies and the entire universe. The *Blazing World* is the culmination and transformation of *Observations upon Experimental Philosophy*. Using her own poetical fancy, Cavendish portrays the monstrous world the experimenters would create if their principles were true, inhabited by hermaphroditical creatures who are foolish and absurd. The nuanced and sophisticated attack on the experimenters of the Royal Society is the print equivalent of Cavendish's visit to the society: ridicule becomes romance.

### THE NEW BLAZING WORLD

*Blazing World* is one of many utopian works produced in early modern Europe. Like Thomas More in *Utopia*, Francis Bacon in the *New Atlantis*, and Cyrano De Bergerac in *Histoire comique contenant les États et Empires de la lune* (1657), Cavendish uses the trope of other worlds to comment on her own. While all these earlier works may be her inspiration, in *Blazing World* Cavendish's vision is far from utopian—upheaval is a constant threat in the Blazing World, and the Empress leads her subjects in wars against other worlds. Her new world differs from other imaginary lands in being ruled by a woman, who is not meant to be laughed at but worshiped. Satire in the Blazing World extends only to the subjects of this imperial lady, not to the monarch herself.

In *Blazing World*, Cavendish takes the monstrous, distorted objects Robert Hooke had displayed with the microscope and turns them into actors who are both the subjects and scientists of her alter ego, the Empress. Rejecting Hooke, Cavendish creates a new world through imagination rather than through micrography.[56] "Art produces most commonly hermaphroditical figures," she argues in *Observations*, and in *Blazing World* Cavendish parodies both the product and producers of unnatural nature.[57]

Of all her works, *Blazing World* attracts the most attention from scholars. Not only does it function as a counterpoise to *Observations upon Experimental Philosophy* (with which it is published), but it also incorporates elements from her natural philosophy, her poetics, her orations, and her plays. It is as much a hybrid as the creatures that populate it, including, besides the beast-men, an amalgamated main character, who at one point combines the souls of Margaret Cavendish and the Duke of Newcastle into her own. In the dedication to the reader at the beginning of *Blazing World*, she states her purpose, "to delight the Reader with variety, which is always pleasing." Like Nature, who delights in variety, Cavendish will create a world composed of many parts, "Romantical . . . Philosophical . . . and Fantastical," although "agreeable to the subject I treated of in the former parts."

Reflecting Cavendish's practice of presenting her ideas in both a serious and a fanciful manner, *Blazing World* can be seen as an extended commentary on *Observations*, but directed at a much broader audience. What philosophy might not accomplish, romance and fantasy will achieve. Consequently, Cavendish proclaimed at the beginning of *Blazing World*, in one of her most famous statements, "Though I cannot be Henry the Fifth, or Charles the Second, yet I endeavour to be Margaret the First; and although I have neither power, nor time nor occasion to conquer the world as Alexander and Caesar did; yet rather then not to be Mistress of one, since Fortune and the Fates would give my none, I have made a World of my own: for which no body, I hope, will blame me, since it is in every ones power to do the like."[58]

The creation of a world is an act of power, and that power will be expressed in *Blazing World* by a continuous and multiplying process of creation and conquest of new worlds by the Empress, and by Cavendish herself—whom Cavendish presents as a character in her own fiction. In the Blazing World, after the heroine marries the Emperor (who at first believed she was a goddess), he gives "her an absolute power to rule and govern all that World as she pleased. But her subjects, who could hardly be perswaded to believe her mortal, tender'd her all the veneration and worship due to a Deity."[59] Unlike the She-Anchoret, who had to die to be worshiped as a goddess, the Empress achieves both deification and power in life.

At first, in the romantic beginning of the story, it did not seem that would be her fate. *Blazing World* tells the story of a young lady who escapes from the importunate desires of a lower-class merchant by traversing the pole between two worlds. Kept alive by the warmth of her beauty while her abductors die in the arctic seas, the maiden encounters "strange Creatures; in shape like Bears, onely they went upright as Men." She is at first stricken with fear, but she is treated with "all civility and kindness imaginable." These monsters apparently possess the politeness the experimental philosophers lack, although later in the text they will prove to be not so gracious. Soon the young lady meets other denizens of this strange world, including fox-men who, like the bear-men, walk upright.[60] In her earliest foray on natural philosophy, Cavendish had argued that the property that made humans human was the ability to walk upright. Apparently, these creatures qualify on that account, as well as possessing language, which the young lady soon learns.

But beast-men transcend the merely human. The beast-men of the Blazing World function in at least three different ways in Cavendish's text. In the first case, they simply parody what Cavendish believed was a ridiculous experimental culture. In the second and more serious case, they represent a civil society that potentially can disturb the peace of the state. Third, and somewhat paradoxically, they demonstrate that experimenters, chemists, and astronomers can function intelligently and usefully if they are governed by a wise and ambitious ruler.

Almost the first thing the Empress does after becoming the absolute ruler of the Blazing World is to form scientific societies. She takes those of her subjects who have a natural aptitude in the arts and sciences, "for they were as ingenious and witty in the invention of profitable and useful Arts, as we are in our world, nay more; and to that end she erected Schools and founded several Societies."[61] Here, through projection into an imaginary world, Cavendish realizes her desire to found schools dedicated to her own philosophy, although first her avatar has to disabuse her natural philosophers of some of their incorrect opinions. The Empress decrees: "The Bear-men were to be her Experimental Philosophers, the Bird-men her Astronomers, the Fly- Worm- and Fish-men her Natural Philosophers, the Ape-men her Chymists, the Satyrs her Galenic Physicians, the Fox-men her Politicians, the Spider- and Lice-men her Mathematicians, the Jackdaw- Magpie- and Parrot-men her Orators and Logicians, the Gyants her Architects, etc."[62]

While the Empress will spend some time with the political beasts, most of her attention is directed at the faux scientists. Later in the text, she tells the Duchess about "the study of Natural Causes and Effects, which was her chief delight and pastime, and that she loved to discourse sometimes with the most Learned persons of that World."[63] Unfortunately, some of the absurdity and uselessness of the investigators of nature Cavendish had decried in *Observations* is also found in their man-beast mimics in the Blazing World. One by one, the monstrous hybrids describe their scientific studies to the Empress, who greets their ideas with mixed parts of incredulity, suspicion, and approbation. She is particularly hard on the bear-men, her experimental philosophers.

Cavendish's choice of bears to represent experimenters demonstrates a kind of cultural canniness—and sense of humor—meant to speak to an audience of readers familiar with the anthropomorphic qualities of animals. Bears had long been an object of the public gaze in early modern Europe: they were displayed and mocked in the bear gardens of Elizabethan and early Stuart England, where they were ripped apart by dogs. Cavendish here implies that both experimenting men and performing bears operate as public entertainment, thus collapsing the cultural pretensions of the experimental philosophers.

First the Empress commands the bear-men to use their telescopes to observe the heavens, "but these Telescopes caused more differences and divisions amongst them, then ever they had before." The experimental philosophers argue about whether the earth or the sun is moving and whether the moon is inhabited and possesses hills and valleys. The Empress had expected her experimental philosophers, at least, to know something about comets or blazing stars—her bird-men astronomers had already told her that in the Blazing World there were "none other but Blazing-stars, and from thence it had the name that it was called the Blazing World."[64] The fact that Cavendish appropriates comets, which were a subject of heated debate in the Royal Society after the appearance of comets in 1664 and 1665, as the designated space for her new world is parodic by itself. But it also emphasizes the foolishness of the experimental philosophers when the bear-men, unable to agree on an explanation for comets, are commanded to break their telescopes, for, as the Empress argues, "your Glasses are false Informers, and instead of discovering the Truth, delude your senses."[65] The bear-men beg her to spare their telescopes because "they "take more delight in Artificial delusions, then in natural truths."[66] In fact, if they broke their instruments, they argue, "we shall want imployments for our senses, and subjects for arguments." If the truths of natural philosophy were known, there would be no controversy, so "we should want the aim and pleasure of our endeavours in confuting and contradicting each other; neither would one man be thought wiser then another, but all would either be alike knowing and wise, or all would be fools."[67]

Clearly, the bear-men are not wise, but nevertheless they might be danger-ous. The Empress allows the bear-men to keep their telescopes, "but upon con-dition, that their disputes and quarrels should remain within their Schools, and cause no factions or disturbances in State, or Government."[68] Experimental philosophy was not only a joke but also a threat to public order, in the Blazing World as much as in the real world. Cavendish will figuratively prevent disor-der, at least in her imaginary creation, when her Empress governs the New Blazing World.

Seeking to placate their ruler, the bear-men decide to show her some of their other instruments, including some microscopes. However, they only dig them-selves in deeper. First they display the fourteen thousand eyes of a drone-fly, an-other specimen Hooke had depicted in *Micrographia*. When the Empress suggests that these eyes might really be pearls and not eyes, the experimenters "smilingly answered her Majesty, That she did not know the vertue of those Microscopes; for they did never delude, but rectifie and inform the senses; nay, the World, said they, would be but blind without them, as it had been in former ages before those Mi-croscopes were invented."[69] Then the bear-men show her a louse and a flea through a microscope; they "appear'd so terrible to her sight, that they had almost put her in a swoon." When she inquired whether "their Microscopes could hinder their biting, or at least shew some means how to avoid them," they answer "that such Arts were mechanical and below the noble study of Microscopical observations."[70]

The bear-men thereby demonstrate both their inanity and their arrogance: they neither can nor want to relieve beggars of lice.

Lice themselves, as we saw in the Introduction in yet another twist, become reified as lice-men who unsuccessfully weigh air, like their real world counterparts. But these insect-men investigators of nature, who endeavor "to measure all things to a hairs breadth, and weigh them to an Atome," Cavendish writes, "seldom agree, especially in the weighing of Air, which they found a task impossible to be done." So much for the careful measurements Robert Boyle did with the air-pump. Cavendish figuratively crushes both lice and the society in one fell swoop by having the Empress tell the lice-men "that there was neither Truth nor Justice in their Profession; and so dissolved their society."[71]

The bear-men and the lice-men practice what Cavendish in *Observations* had called "the emulating Ape of Nature."[72] This is an analogy Cavendish had borrowed from Van Helmont; in *Philosophical Letters*, after remarking that "Art is but a Particular effect of Nature, and as it were, Nature's Mimick or Fool," she adds that Van Helmont "confesses it himself, when he calls the *Art of Chymistry, Nature's emulating Ape*."[73] Indeed, Cavendish may learned this metaphor from Robert Boyle, who in *The Sceptical Chymist* claimed that Paracelsians "are like Apes, if they have some appearance of being rational, are blemish'd with some absurdity or other, that when they are Attentively considered, makes them appear ridiculous."[74]

In early modern Europe, apes were both circus performers (fools) and imitators of man (mimics). They both fascinated and appalled by their close physical resemblance to humans and the tricks they could perform.[75] Cavendish herself, in describing the delights of novelty in *Philosophical Letters*, wrote, "I have seen an Ape, drest like a Cavalier, and riding on Horse-back with his sword by his side, draw a far greater multitude of People after him, then a Loadstone of the same bigness of the Ape would have drawn Iron; and as the Ape turn'd, so did the People, just like as the Needle turns to the North."[76]

Apes are archetypical hybrid figures, *stupores mundi*. So are the ape-men of the Blazing World, but they do not generate wonder. While her ape-men are certainly fools, they lack the entertainment value even of magnets: they are boring, and after their "tedious" discussions of the principles of nature the Empress, "being so much tired that she was not able to hear them any longer, imposed a general silence upon them."[77] She decides to fill the silence by teaching them vitalistic materialism and ordering them not to "waste your time in such fruitless attempts, but be wiser hereafter; and busie yourselves with such Experiments as may be beneficial to the publick."[78]

The only natural philosophers the Empress truly admires are the worm-men, who are intelligent enough to endorse the materialism familiar to anyone who has read Cavendish's scientific treatises. When the Empress questions them about forms, they answer "that they did not understand what she meant by this expression; For, said they, there is no beginning in Nature, no not of Particulars, by reason Nature is Eternal and Infinite, and her particulars are subject to infinite changes and transmutations by vertue of their own corporeal, figurative self-motions; so that there is nothing new in Nature, nor properly a beginning of any thing."[79]

Why does Cavendish give worms, seemingly the lowest animal on the great chain of being, the glory of being her spokesmen? In her 1657 autobiography Cavendish analogized herself to a kind of worm, albeit a useful sort of worm. "I must say this in the behalf of my thoughts, that I never found them idle; for if the senses bring no work in, they will work of themselves, like silk-wormes that spins out of their own bowels."[80] Moreover, as we saw before, Cavendish argued, women become bestial and irrational when denied education: "They become like worms that onely live in the dull earth of ignorance."[81] In Blazing World, the image of the unknowing worm (and woman) is transposed into that of the most prescient of animals, living in a place of teeming life. It is all a question of perception, which it turns out the worm-men have in abundance. When the Empress commands the bear-men to lend their microscopes to the worm-men in order to see better, the rude experimentalists reply "that their Glasses would do them but little service in the bowels of the Earth, because there was no light." To which the worm-men answer "that although they could not say much of refractions, reflections, inflections, and the like; yet were they not blind, even in the bowels of the Earth; for they could see the several sorts of Minerals, as also minute Animals, that lived there, which minute animal Creatures were not blind neither, but had some kind of sensitive perception that was as serviceable to them, as sight, taste, smell, touch, hearing, etc was to other animal Creatures."[82] Worms, thus, reveal "that Nature has been as bountiful to those Creatures that live underground . . . as to those that live upon the surface of the Earth, or in the Air, or in Water." They also reveal that natural perception is not dependent on external senses or external instruments. Knowledge is interior and can be generated inside the earth or inside the mind.

In early modern Europe, worms would have a particular advantage in understanding the nature of material being. Almost everyone, in elite and popular culture alike, believed that worms were produced by spontaneous generation. In Observations, Cavendish argued, "But such insects, as Maggots, and several sorts of Worms and Flies, and the like, which have no Generator of their own kind, but are bred out of Cheese, Earth and Dung, etc. their Production is onely by Metamorphosing, and not Translation of Parts."[83] Until the late seventeenth- and early eighteenth-century experiments of Redi, Leeuwenhoek, Swammerdam, and Malpighi proved otherwise, everyone believed that decaying flesh generated worms and flies. Historian Keith Thomas remarks, "This demonstration [of the microscopy] came as a great relief to the pious, for whom spontaneous generation threatened to make a Creator unnecessary."[84] Cavendish, of course, thought that all creation contained its own principle of self-moving matter, including worms. In *Philosophical Letters*, she argued, "When a Worm is cut into two or three parts, we see there is sensitive life and motion in every part, for every part will strive and endeavour to meet and joyn again to make up the whole body." Elsewhere in the text, citing her own observation of tiny worms, she states, "Yet they were more agil and fuller of life, then many a creature of a bigger size . . . and I do verily believe that these small creatures may be great in comparison to others which may be in nature."[85]

These lively worms are yet another expression of the minute elements of nature, which include the very tiniest—the atomic fairies. They are the prototype for the all-knowing matter that composes Cavendish's material nature. It comes as no surprise, therefore, that the worm-men know more about the workings of nature than any other of the investigators of nature in the Blazing World. In a sense, they are nature and Cavendish's material principles metamorphosed into a specific analogical form. In *Blazing World*, the Empress applauds the materialism of the worm-men, whose views come closest to Cavendish's own philosophy: "The Empress was so wonderfully taken with this discourse of the Worm-men, that she . . . yielded a full assent to their opinion, which she thought the most rational that ever she had heard yet."[86] Paradoxically, worms, the lowest form of animal life, become the most prescient of natural philosophers. Their ascendancy, overturning and undermining the established hierarchies of nature and knowledge, validates the role Cavendish had fashioned for herself: a female natural philosopher—also a hybrid—who knew more about the earth than any of her male fellows.

Worm-men represent a culmination of the natural order, even in their monstrosity. Cavendish contended that these creatures, at least, might be better natural philosophers than their human examiners because they know the nature of their kind better than any observer could. This view is in direct contradistinction to that of Robert Hooke, who in *Micrographia* argued that God should be admired for the diversity of his creation but that "we should leave off to admire the creature, or to wonder at the strange kind of acting in several Animals, which seem to savour so much of reason; it seeming to me most manifest, that those are but acting according to their structures, and such operations as such bodies, so compos'd, must necessarily, when there are such and such circumstances concurring, perform."[87]

Hooke, in this argument, adopted the Cartesian conclusion that animals are automatons devoid of reason and will. As we saw in Chapter 6, Cavendish had already challenged this assertion in *Philosophical Letters*. Cavendish believed that the Cartesian argument was not only wrong but presumptuous. In *Observations*, the object of her ridicule is obvious: "But Man, out of self-love, and conceited pride, because he thinks himself the chief of all Creatures, and that all the World is made for his sake; doth also imagine that all other Creatures are ignorant, dull stupid, senseless and irrational, and he onely wise, knowing and understanding."[88]

Ultimately, however, the rationality of the beast-men leads to the same kind of divisions Cavendish experienced in her own world. The Empress will not allow civil war and faction to haunt her created world, even if some of her subjects are decent natural philosophers. The Empress tells Cavendish, who has appeared as a character in her own fiction, that the "continual contentions and divisions between the Worm-, Bear-, and Fly-men, the Ape-men, the Satyrs, the Spider-men, and all others of such sorts, that I fear they'l break out into open Rebellion, and cause a great disorder and ruine of the Government." Cavendish advises the Empress "to dissolve all their societies" and reinstate an absolute sovereignty.[89] She does so, and in the second part of *Blazing World*, these contentious subjects are transformed into her messengers and soldiers.[90]

The beast-men become useful citizens of the state when each uses the knowledge of its kind to aid the Empress in her conquests. The second part of *Blazing World* consists of a description of the Empress's war, advised by Margaret Cavendish, on the nations fighting her own native land, the country of E S F I. Before the battle, she announces her intention to protect her country from its enemies, and the citizens "all kneeled down before her, and worshipped her with all submission and reverence." All they have to do to earn her aid, she instructs, is to acknowledge her "Power, Love and Loyalty to my Native Country; for although I am now a great and absolute Princess and Emperess of a whole World, yet I acknowlede that once I was a Subject of this Kingdom." Much like the audience of her books, the citizens of E S F I are astounded: "Some said she was an Angel; others, she was a Sorceress; some believed her a Goddess; others said the Devil deluded them in the shape of a fine Lady."[91] Cavendish knew that she made a splash wherever she went, in real or fictional spaces.

The Empress orders her bear-men to use their telescopes to spy out her enemy's towns and cities. The worm-men burrow through the earth and place firestones under the foundations of towns. The fish-men provide transportation, so the Empress seems to be walking on water. These and similar manifestations cause the spectators to admire her as "an uncreated Goddess," a role Cavendish attempted to imitate in her visit to the Royal Society, to somewhat different effect.[92]

In the Blazing World, experimental philosophers serve and respect their ruler, and instead of playing with toys, they use their instruments to the greater glory of their country and its people. Ultimately, Cavendish was making an argument about power and presumption in Observations and Blazing World. The arrogance of the man-beasts, and possible factions they might cause, are avoided when they submit to and obey their transcendent female ruler. The Empress, having manifested her world-creating and world-destroying power, retires to devote her time to the study of "natural Causes and Effects, her chief delight and pastime." She has achieved the status and honor that the experimental philosophers sought, and she has preempted the control over nature they claimed.

But even in the Blazing World, the character of Cavendish cannot realize all her ambitions—she is an adviser but not a ruler of any world. In the narrative, the Duchess bemoans the fact that she is not "Emperess of a World, and I shall never be at quiet until I can be one."[93] Seeking to help her, the Empress calls "the immaterial spirits," who wonder why the Duchess is so despondent. After all, she can make any sort of immaterial world, whether out of atoms or whirlpools or pressures and reactions, so "What need you to venture life, reputation and tranquility, to conquer a gross material world?"[94] Taking their advice, the Duchess decides to "reject and despise all the worlds without me, and create a world of my own." She first attempts to make a world out of atoms, which creates such dust in her mind that she can't even think. A Cartesian world makes her faint and dizzy, while a Hobbesian universe would give her a terrible headache.[95] The obvious answer is to create a world out of the material principles of Cavendish's philosophy. Once completed, the Duchess invites the Empress "to observe the frame, order, and Government of it. Her Majesty [the Empress] was so ravished with the perception of it, that her soul desired to live in the Duchess's World; but the Duchess advised her to make such another World in her own mind; for, said she, your Majesties mind is full of rational corporeal motions, and the rational motions of my mind shall assist you by the help of sensitive expressions, with the best instructions they are able to give you."[96]

Multiplying imaginary worlds are the vehicles Cavendish uses for realizing her own fantasies. They are real in the sense that Cavendish believed that all rational conceptions, including fantasies, are composed of the same material bodies. At the end of *Blazing World*, Cavendish invites her readers to become her subjects and, if they do not want to, to create worlds of their own. "But," she warns them, "let them have a care, not to prove unjust Usurpers, and to rob me of mine; for concerning the Philosophical World, I am Emperess of it my self."[97]

When Cavendish attended the meeting of the Royal Society, the year after *Observations upon Experimental Philosophy* and *Blazing World* were published, she swept into Gresham College the way the young lady swept into the Blazing World. The experimenters did not know it, but they were being turned into the subjects of her majesty, the creator of words and worlds. But her appearance was not meant to honor Hooke, or Boyle, or any of their compatriots. She neither desired the patronage of the society nor wanted to give it her support. Rather, she was translating into the actual world what her doppelganger had already accomplished in her imaginary world. She was revealing them to be fools and charlatans creating artificial monstrosities, absurd in themselves and in what they did. Cavendish was no fool herself but rather a woman with a plan—to establish her own philosophy and subvert her rivals. While they played with toys and baubles, she would reign over the natural philosophic universe and any other worlds she happened to create with her rational conceptions. In her last work, *Grounds of Natural Philosophy* (1668), she culminated her exploration of strange new worlds and strange new beings.

# chapter eight

# Material Regenerations

*N*ever one to leave a fertile image or a speculative fancy undeveloped, Cavendish filled her last original works, the 1668 edition of *Plays* and *Grounds of Natural Philosophy* (1668), with elaborate themes and images from her earlier work. But the last two differ from the earlier treatises (and fancies) in tone. By this time in her career, having fully developed her own natural philosophy and delivered her scathing critiques of other philosophies—and having lampooned the experimenters in print and person—Cavendish was ready for a good laugh at the expense of her peers, and even herself. These works show a willingness to explore the often bizarre implications of her vitalistic materialism without fear of any kind of authority, either philosophic or religious. *Grounds of Natural Philosophy* features speculations about other worlds, other places, and other kinds of humanity and questions Christian doctrines of the soul and resurrection. But these speculations display humor and even whimsy, largely absent from her earlier philosophic texts, recalling *Poems, and Fancies* or *Natures Pictures*, her more lighthearted works of the 1650s.

The works of 1668, which include a new set of plays as well as her last philosophic treatise, share a preoccupation with gendered themes. Cavendish's most famous play, "The Convent of Pleasure," pictures a kind of heavenly female retreat where women can enjoy good food, sweet smells, and each other's companionship without the disturbing presence of men. Such a life is even better than the imagined power of an absolute queen. One character remarks, "I had rather be one in the *Convent of Pleasure*, then Empress of the whole World."[1] The heroine of this play is Lady Happy, whose vision of a perfectly pleasurable world echoes the description of a happy world in *Grounds of Natural Philosophy*—where the material parts are regular, know each other perfectly, and do not fear the dissolution of death because they know their parts will be united with other parts. Indeed, "all Human Creatures of that World, are so pleasant and delightful to each other, as to cause a general Happiness."[2] The happiness in the continuation of species pales, however, in comparison with the restoration of individual beings to life, which Cavendish suggests could be achieved with a sort of restoring bed or womb.

Thus, as in her earlier works, different genres comment upon each other. They are intertwined, like Cavendish's material principles. The *Grounds of Natural Philosophy* and the 1668 edition of *Plays* are also more humorous.[3] Instead of tales of adventure and female empowerment, such as in the 1662 edition of *Plays*, most of these plays fit into the genre of Restoration comedy, with the dominant theme being the search of dissolute young men for rich wives (or easy conquests) and women's quests for rich husbands. They are similar to the plays written by her husband, who even contributed scenes to "The Presence."[4] These later plays display a continued sensitivity on Cavendish's part to her image in society, but they also mock some of her own poetic and philosophic conceits. Cavendish appropriates the newly popular form of the comedy of manners as yet another way of presenting herself, but in this case she is implicitly arguing that the laughter is with her, not against her.

The year 1668 also saw the publication of a new edition of *Blazing World* in a stand-alone form, unattached to *Observations*.[5] This is Cavendish's only work, besides *Poems, and Fancies*, dedicated to women. Both are inscribed "To all Noble and Worthy Ladies," and both note that most women are not interested in the serious contemplation of nature. "For I have observ'd," writes Cavendish in 1653, "that their Braines work usually in a Fantasticall motion," while in 1668 Cavendish comments, "Most Ladies take no delight in Philosophical Arguments." Cavendish here seems to be responding to the general lack of enthusiasm her works had received. But since women—because of lack of education—are more inclined to enjoy fancies, she will once again try to engage them in her natural philosophy by way of fantasy. And so, she adds, "And if (Noble Ladies) you should chance to take pleasure in reading these Fancies, I shall account my self a Happy Creatoress: If not, I must be content to live a Melancholly Life in my own World."[6]

By 1668, Cavendish had retired to Welbeck Abbey, the Newcastle estate in Nottinghamshire. Her husband was no longer a favorite of the court, and it appears that Cavendish's 1667 biography of him, *The Life of the Thrice Noble, High and Puissant Prince William Cavendish, Duke, Marquess, and Earl of Newcastle*, did not bring him the recognition Cavendish passionately believed he deserved. Pepys's reaction to it also shows that Cavendish's visit to the Royal Society, whatever its aim, had done nothing to alleviate her reputation for eccentricity. The diarist wrote in March 1688, I "stayed at home reading the ridiculous history of my Lord Newcastle wrote by his wife, which shows her to be a mad, conceited, ridiculous woman, and he an ass to suffer her to write what she writes to him and of him."[7]

Pepys's accusation that Cavendish was mad was often repeated in the centuries to follow. Centuries of challenges to the duchess's sanity cloud the legitimacy of her claim to be a natural philosopher. They also shadow the questions of whether Cavendish was a feminist before her time and what it meant to develop a gendered natural philosophy in the seventeenth century. Some interpreters see Cavendish's natural philosophy as a way to broaden the discussion of the development of science during the seventeenth century to include a woman. Others question Cavendish's "feminism" because so much of her work seems either antithetical to or unaffected by gender concerns.[8]

Perhaps the key to Cavendish's disputed feminism is to understand her as she understood herself. She recognized that she was doing something very unusual for a woman of her time. She identified male presumption as the enemy of both Nature and women and male dominance as a threat to all living creatures. Her emphasis on the power of nature, and her depiction of powerful women, formed a self-conscious reply to the subservience men demanded of women in early modern Europe. Cavendish was aware of her sex, personally and collectively, and this recognition produced the gendered elements of her natural philosophy.

## THE RETURN OF THE BEAST-MEN: THE 1668 PLAYS

Recent scholarship on Cavendish's plays concentrates on their use of social satire, what they reveal about gender in the seventeenth century, and their place in the theatre of the time.[9] Cavendish's earlier plays often included heroic women, who clearly reflected the grandeur with which the duchess endowed her self-projections in *Natures Pictures* and later in *Blazing World*. Later plays also featured representations of Cavendish. In "The Presence," a character named Lady Bashful is thought a fool but is recognized as a witty and intelligent woman by Lord Loyalty, who then marries her. Another projection might be Lady Happy in "The Convent of Pleasure," but the clearest avatar is Lady Phoenix, in "A Piece of a Play." Lady Phoenix presents herself with such splendor that she "will astonish all her Spectators." She "is clothed all with light, and the beams issuing from that light, makes her train many miles long, which is held up by the Planets." Here indeed is a spectacle to behold, just as Cavendish herself was in 1667 and the Empress was in *Blazing World*. The character's plumage parallels that of her eponymous prototype. But some characters in this play, like Pepys in real life, were not impressed by Lady Phoenix, including the bluntly named Mr. Ass, who describes her "as proud as Lucifer." More knowingly, Lady Buzzard indicates that Lady Phoenix "feeds only upon Thoughts," while Mrs. Dormouse's description of her makes clear Phoenix's kinship with Cavendish: "She is of a studious nature, in a retired life, ever retireing from much Company, and of a careless humour, not regarding what the World says, or doth; in Company she is of a free Disposition, and an airy Conversation; she is civil to strangers, kind to acquaintances, bountiful to her servants, and charitable to the poor; also she is humble to those that are respect-ful, but severe to those that are rude."[10]

The mythic phoenix's most important characteristic is its repeated death and resurrection, as well as its splendid decoration. Just as the phoenix dies in the flames and is reborn alive, Cavendish may perish in the calumnies of her contem-poraries, only to be reborn in future ages to the fame she desired above all else.

Thus animal imagery, so essential to her natural philosophic works, also in-forms her drama. "A Piece of a Play" is essentially a comedy of manners, with the requisite fops, intriguers, fools, and clever virgins. What is remarkable about this text is the fact that Cavendish had once intended it for the delectation of the Empress and the Duchess. At the beginning of the play, she tells her readers that she had intended it to be part of *Blazing World* but ultimately did not complete it, "finding that my Genius did not tend that way."[11] Perhaps Cavendish felt the ruler of the Blazing World and her companion might have been discomforted by the play's irreverence toward the imaginary world's inhabitants. The most un-usual feature of "A Piece of a Play" is the naming of the characters—here we meet again with a worm-man, a fox-man, and a bear-man. But instead of being reified experimental philosophers or their objects of study, these beast-men are the social types of Restoration court society. Cavendish is using a joke to make another joke. Perhaps the most humorous figure of the collection is Sir Puppy Dog-man, who unfortunately did not figure in *Blazing World* but here is taught how to be a "mode-wit" by Mr. Ass-man, also a no-show in the earlier romance. Lord Bear-man, who seeks the same social skill, prepares to woo Lady Monkey by saying,

> I do confess I want those rules and arts
> As such Men have that are nam'd Men of Parts.
> But such Men as these are not natural,
> For all Mode-Gallants are artificial;
> But for your sake, I will go to Mode's School
> To learn Mode's fashions for to play the Fool.[12]

This bear-man is not yet a natural philosopher as he is in *Blazing World*—he still has to learn how to play with artificial arts and rules, which will make him a fool.

In the play, mode-gallants take the place of experimentalists, but they share the silliness and pretensions of their parodied forebears.

Dandies and experimenters may be fools, but comedy gave Cavendish the opportunity to portray a real fool. In "The Presence," a Fool, for no particular dramatic purpose, describes a vision he has seen of peculiar beings. "Oh ladies," he says. "Oh the strange sights that I have seen! The monstrous strange sights that I have seen!" Asked to explain, the Fool describes his dream: "I saw Men with strange Heads, and as strange Bodies; for they had the speech of Men, and the upright shape of Men, and yet were partly like as other Creatures; for one Man had an Asses head, and his body was like a Goose; another Man had a Jack-a-napes-head [monkey], but all his body was like a Baboon, and he shew'd tricks, as Jack-a-napes and Baboons use to do; another Man had a Swines head, and all his body was like a Goat; Another had a head like a Stag, with a large pair of branched Horns, and all his body was featur'd like a Woodcock, and his arms were feather'd as a Woodcocks wings, but he could not fly from his disgrace, for his Horned head did hinder the flight of his Wings."[13] But these beast-men are singular in having monstrous sisters, members of a sex never described in *Blazing World:* "Then I saw a Woman that was not like a Mare-Maid, for Mare-Maids are like Women from the head to the waste, and from the waste like a Fish; but this Woman was like a Fish from the head to the waste, and from the waste like a Beast; so that she was a Batons rompus; Another Woman had the eyes of a Crocodile, but her body was like a changeable Cameleon; and many other Monstrous Creatures did I see."[14]

Perhaps Cavendish broadened the parameters of possible monstrosities because her audience in the plays was women or because the Fool was speaking to ladies. The association of women with fish is a commonplace of European folklore, but the duchess is careful to distinguish the Fool's beast-fish-woman from mermaids. Nevertheless, this may be a covert attack on Cavendish's female detractors: in the sixteenth and seventeenth centuries, prostitutes were sometimes referred to as mermaids.[15] The crocodile image is easier to decipher: the crocodile, according to the *Oxford English Dictionary,* "was fabulously said to weep, either to allure a man for the purpose of devouring him, or while (or after) devouring him."[16] Chameleons, of course, were associated with rapid and sycophantic change. John Dryden, a client of William Newcastle's, wrote, "The thin Camelion, fed with Air, receive The colour of the Thing to which he cleaves."[17] The hybrid natures of these female characters emphasize negative qualities associated with women: they are lusty, fickle, rapacious, dangerous, and bestial. Fools are often privileged to speak the truth—this Fool may be voicing Cavendish's view of most of her audience.

Of all the plays published in 1668, "The Convent of Pleasure" has received the most scholarly attention. Its depiction of a kind of feminist paradise, created by the heiress Lady Happy, seems to express the unrealized hope of escape from marriage Cavendish so often expressed. In the play, Lady Happy decides to cloister herself, "since there is so much folly, vanity, and falsehood in Men, why should Women trouble and vex themselves for their sake; for retiredness bars the life from nothing else but Men."[18] Lady Happy, like Lady Bashful, wants to retreat from the world. Like the She-Anchoret, she wants to have nothing to do with men. However, the convent she creates is more like a single-gendered Rabelaisian Abbey of Thélème than an anchorite's cave; here the aim is to live a life of pleasure, eating the finest foods and wearing the most luxurious clothes, and, most important, not having to marry. The description of the Convent of Pleasure replicates the description of the saved in paradise in Cavendish's *Orations* (1662), in which she described "their senses more Perfect and their Appetites more Quick . . . these Glorified Bodies shall have their Senses fill'd and their Appetites Satisfied in a Spiritual manner." The sight will "have the most Beautiful, Splendourous, Pleasant, and Glorious Objects," hearing will be filled with "Harmonious Musick, Melodious Voices and Pleasing Vocal Sounds" as well as "Eloquent Language, Witty Expressions, and Fancy, Exprest both in Verse and Prose"— perhaps the luckiest ones heard Cavendish's works.[19] Likewise, in "The Convent of Pleasure," according to Lady Happy,

> For every Sense shall Pleasure take,
> And all our Lives shall merry make . . .
> Wee'l please our Sight with Pictures rare;
> Our Nostrils with perfumed Air.
> Our Ears with melodious Sound,
> Whose Substance can be no where found;
> Our Tast with sweet delicious Meat. . . .
> Thus will in Pleasure's Convent, I
> Live with delight, and with it die.[20]

Heaven and the Convent do differ in one major way. The pleasures of Heaven are internalized perceptions whose substantial reality comes from being in the mind, not the body. The orator explains, "Thus every Sense shall be Satisfied in a Spiritual way, without a Gross Corporeal Substance, and the Blessed Souls of these Glorified Bodies, and Spiritual Satisfaction of Glorified Senses and Appetites, shall be fill'd with all Perfection . . . and all the Passions Regulated and Govern'd as they ought to be."[21] Such perfection is impossible in the world Lady

Happy creates; irregularity enters in the person of the Prince who dresses as a Princess in order to infiltrate the Convent and marry Lady Happy. In *Grounds of Natural Philosophy* Cavendish argues that an absolutely regular world can be found only in Heaven or on another earth, different from the "Purgatory World" we now inhabit.

Still, the imperfect world Lady Happy creates is full of pleasure because pleasure is natural. She disputes the asceticism of traditional religious retreats because the gods would not require such exactions from humans; she asks, "What profit or pleasure can it be to the gods to have Men to lie uneasily on the hard ground, unless the gods and Nature were at variance, strife and wars, as if what is displeasing unto Nature, were pleasing to the gods, and to be enemies to her, were to be friends to them."[22] In Cavendish's philosophic works Nature and God are usually allies, not enemies. Nature and the gods share the same moral universe in "The Convent of Pleasure": they both want people to have the greatest amount of pleasure possible and the least amount of pain. The Epicurean calculus of pleasure and pain is the founding principle of Lady Happy's paradise: "Wherefore, if the gods be cruel, I will serve Nature; but the gods are bountiful, and give all, that's good, and bid us freely please our selves in that which is best for us: and that is best, what is most temperately used, and longest may be enjoyed, for excess dost wast it self, and all it feeds upon." According to the abbess of this most unusual convent, most people pray only when forced to do so by adversity, and they do not deserve the "happiness of ease, peace, freedom, plenty and tranquility in this World, nor the glory and blessedness of the next."[23]

*Grounds of Natural Philosophy* reflects the discussion of pleasure and pain that preoccupies Lady Happy. In this work Cavendish imagined another well-ordered and happy world, which, like the Convent of Pleasure, is full of like-minded fellows. She also contemplated the nature of a miserable world, whose inhabitants are mixed beings, partly one animal and partly another. Sharing the hermaphroditical character of so many of her works, *Grounds of Natural Philosophy* also tackles the question of whether human beings can be regenerated. Just as the 1668 edition of *Plays* poked sly fun at many of the conceits Cavendish had developed in her earlier works, Cavendish's last philosophic work plays with many of the themes that pulse through her works.

## WORLDS AND WOMBS

*Grounds of Natural Philosophy* is the third incarnation of *Philosophical and Physical Opinions*, itself an elaboration of the 1653 *Philosophicall Fancies*, which was reprinted in the later text; *Philosophical and Physical Opinions* was published in 1655 and 1663. Cavendish was never satisfied with the earlier editions, which were supposed to give a crystal-clear rendition of her natural philosophy. In *Observations* she wrote, "I do ingeniously confess, that both for want of learning and reading Philosophical Authors, I have not expressed my self in my Philosophical Works, especially in my Philosophical and Physical Opinions, so clearly and plainly as I might have done."[24] Hence, all her later works are filled with clarifications of one sort or another, which develop aspects of her material philosophy. *Grounds of Natural Philosophy* does much of the same, and many parts of the text are simply reproductions of her earlier work in somewhat shortened form. The only change Cavendish notes herself is to credit Nature with an even more extensive self-knowledge and power than she had possessed in earlier descriptions. Her "Infiniteness" no longer limits her "Absoluteness."[25]

The other, more obvious transformation in this text is the elimination of most of the prefatory material that had introduced her other works, including the many prefaces and dedications. *Grounds of Natural Philosophy* has only one dedication, "To the Universities of Europe," another display of Cavendish's patronage scheme to gain admittance to the scientific community. But this text does include an unusual apologia, or at least a kind of plea for her work, which echoes a theme largely absent from her work since *Poems, and Fancies*. In the earlier text, she begged, "Condemne me not for making such a coyle/About my Book, alas it is my Childe."[26] In 1668, she wrote, "And it is my confidence, That you will be propitious to the Birth of this beloved Child of my Brain," which she never put "to suck at the Breast of some Learned Nurse" and instead decided "Obstinately . . . suckle it my self, and bring it up alone, without the help of any Scholar."[27] This adoption of the role of single mother for her child/work frames *Grounds of Natural Philosophy* as a gendered creation. And, in fact, the idea of procreation and creation, as well as regeneration and resurrection, will permeate this book. As in so many of her other works, Cavendish added a fantastical addendum to her serious natural philosophy, which in this case includes the possible material regeneration of corpses. She hoped for a similar fate for her work, which, she says, the universities could grant "everlasting Life," but if it is found wanting, it should "be not buried in the hard and Rocky Grave of your Displeasure; but be suffer'd, by your gentle silence, to lye still in the soft and easie Bed of Oblivion."[28]

*Grounds of Natural Philosophy* begins by restating Cavendish's holistic materialism; there are three parts of matter, "the Rational Parts, the Sensitive Parts, and the Inanimate Parts; which three sorts of Parts are so join'd, that they are but as one Body; for, it is impossible that those three sorts of Parts should subsist single,

by reason Nature is but one united material Body."[29] As usual, rational and sensitive matter is self-moving and vital, and the existence of matter precludes the existence of a vacuum. This triune matter comprises all being—animal, vegetable,
mineral, and elemental—and it is substantial. Cavendish here again denies the
existence of incorporeal substance. "But as for Matter," she argued, "there may
be degrees, as, more pure, or less pure; but there cannot be any Substances in
Nature, that are between Body, and no Body."[30] Moreover, it is impossible that an
incorporeal substance could begin the motion that characterizes the universe.
Rather, motion is eternally part of an eternally moving material universe; this matter is never annihilated but only dissolved and recomposed. Thus, vitalistic materialism underlies a universe where creatures can be recomposed from their substantial parts, just as Cavendish is restoring her natural philosophy in a more
cogent form—although she admits that is unlikely that any body would be exactly
the same as before its recomposition.[31]

Cavendish reiterates the epistemological skepticism that characterized her
attack on the natural philosophers and the experimental philosophers. Human
perception itself is limited to only the "Exterior Parts" of another creature and
cannot penetrate into its internal being, not even with microscopes or telescopes.
The motions that create "Vegetables, Minerals, Elements, and the like," she argued, "the subtilest Philosopher, or Chymist, in Nature, can never perceive, or
find out."[32] As in her earlier work, this skepticism becomes a license to explore
a world of possibilities. "A Man," she states, "may suppose or imagine what the
innate nature of such a Vegetable, or Mineral, or Element is; and may imagine or
suppose the Moon to be another World, and that all the fixed Starrs are Sunns;
which suppositions, Man names Conjectures."[33]

The exploration of the idea of other worlds becomes the basic motif of the appendix to *Grounds of Natural Philosophy*, the only completely new part of this last
work. The duchess had long been preoccupied by the possibility of other kinds of
worlds. In *Poems, and Fancies*, Cavendish speculated about the existence of minute subatomic lands: "What severall Worlds might in an Eare-ring bee. . . . /And
if thus small, then Ladies well may weare /A World of Worlds, as Pendents in each
Eare."[34] In *Philosophicall Fancies*, Cavendish argued, "Thus may another World
though matter still the same. /By changing shapes, change, Humours, properties,
and names"; this world could be inhabited by flower ladies and metal men.[35] *Natures Pictures* is awash with other worlds, in the air and in the earth, and the 1655
edition of *Philosophical and Physical Opinions* embraced the possibility of alien
creations: "As the Sun differs from the earth and the rest of the planets, and earth
differs from the seas, and seas from the airy skie, so other worlds differ from this

world, and the creatures therein, on different degrees of innate matter . . . so may worlds differ for all we know."[36] Cavendish repeated this idea in *Grounds of Natural Philosophy*. "And certainly," she argued, "there may be, in Nature, other Worlds as full of varieties, and as glorious and beautiful as this World . . . as also, more full of variety than this World, and yet be quite different in all kinds and sorts, from this World."[37]

The imagining of other worlds allows Cavendish to explore the possibly heretical themes she had suggested in her earlier works. When her discussion becomes more problematic from a religious point of view, she abandons the declarative style used in most of the text for dialogue, which allows the parts of her mind to argue like characters in one of her plays or orators at a debate. The duchess had used the rhetorical device of a conversation between the major and minor parts of her mind before, but in the appendix to *Grounds of Natural Philosophy*, it becomes the vehicle for some very strange notions indeed.

In the appendix, as in her earlier works, we find the suggestions that the soul is material and that, if it is sinful, it suffers actual physical torment after death. Likewise, the angels and devils are material, and God himself is in some sense material. In an unusually candid but extraordinarily convoluted way, Cavendish tries to reconcile materialism and Christianity. She writes, "But, considering that Hell and Heaven is described to be Material, it is probable, Spirits are also Material: nay, our blessed Saviour Christ, who is in Heaven, with God the Father, hath a Material Body; and in that Body will come attended by all the Hosts of Heaven, to judg the quick and the dead; which quick and dead, are the Material Parts of Nature: which could not be actually judged and punished, but by a Material Body, as Christ hath."[38] Cavendish quickly adds that she means Christ in this case is judging only "Material Actions," and he does make "Divine and Immaterial Degrees." Nevertheless, the implication is clear: this is a very Nestorian kind of Christ.

Thus, Cavendish continues, if some part of the soul survives death, that part is material and must experience a material Heaven and Hell. Otherwise, the creature would not experience the pleasure and pain resulting from virtue or sin. But if a general resurrection of bodies occurs, Cavendish questions whether all the former parts of a creature could be reunited. If all the parts come together, the resultant reassembled man would be a giant—or if he died young, a dwarf. Or, the Minor Parts argue, he would be "a Monstrous Creature, as having more Parts than was agreeable to the nature of his Kind." To which the Major Parts reply, "The MAN, would be a Society of greater Magnitude; yet, not in any ways different from the Nature of his Kind."[39] The Major Parts usually express Cavendish's own view in this book; a resurrected human being thus would not be monstrous, but he would be big—really big.

And if all individuated beings were resurrected, the earth would dissolve because all its parts that came from the dissolution of human bodies over the long span of time would disappear. Here is a materialist consideration of the last days indeed: What does God do with all those big bodies after the matter of the earth is integrated into its former inhabitants? The answer is equally materialistic; he would command "His Servant Nature, to compose other Worlds for them, into which Worlds they should be separated; the Good should go into a Blessed World; the Bad, into a Cursed World: and the Sacred Scripture declares, That there shall be a New Heaven, and a New Earth."[40]

Paradise and Hell, therefore, are the corporeal destination for resurrected bodies. Somewhere out in space, reward or punishment awaits us all. But Cavendish's sketch of this materialistic eschatology, which can to some extent be reconciled to an orthodox Christian view of salvation and damnation, often segues into a more suspect description of Heaven and Hell.[41] The ultimate destinations for humankind are reconfigured according to the specifics of Cavendish's natural philosophy. The parts of the mind wonder "if there might be a World composed only of Irregularities; and another, only of Regularities."[42] Earlier, Cavendish had defined sin as disobedience to Mosaic Law: "By reason Nature is as much Irregular, as Regular, Human Notions are also Irregular, as much as Regular . . . that occasions Irregular Devotions, and is the cause of SIN."[43] The idea of sin being caused by irregularities in Nature appeared first in *Natures Pictures* and is here used to ground the argument that there might be two kinds of future worlds. Thus, after much discussion, the parts agree, "There might be Regular and Irregular Worlds; the one sort to be such happy Worlds, as they might be named Blessed Worlds; the other so miserable Worlds, as might be named Cursed Worlds."[44] However, the rewards and punishments awaiting individuals are not the ones familiar to us from Dante's *Divine Comedy*. Rather, they reflect the duchess's materialist theory of perception. She writes "that both the Sensitive and Rational Parts of those that are restored to Life, should move in variety of Perceptions, or Conceptions, without variety of Objects: and, that those Creatures (viz. Human Creatures) that are raised from Death to Life, should subsist without any Forrein Matter, but should be always the same in Body and Mind, without any Traffick, Egress, or Regress of Forrein Parts."[45]

This conception of Heaven and Hell expands the versions found in *Orations*, which also colored "The Convent of Pleasure." Each person after death is raised in a material body, but this body is unaffected by sense perception. Instead, each person dwells with his or her own conceptions, the mind's "Visions, or Imaginations," like those on earth. Each individual retains the physical memory of past thoughts, of which "some have been Pleasing and Delightful; others, Displeasing and Dreadful." We are literally haunted by our former conceptions. This is a material Heaven and Hell because, according to Cavendish, even our thoughts and visions are substantial. If we feel tormented, we are tormented. If we are delighted, that too is a material manifestation of the mind.

Having ventured this far into the domain of the blessed and the damned, Cavendish immediately withdraws into the landscape of fancy. In the third part of the appendix to *Grounds of Natural Philosophy*, Cavendish tells the reader that she is going to continue her speculations about happy and miserable, or regular and irregular, worlds. "But," she adds, "pray mistake not these Arguments; for they are not Arguments of such Worlds as are for the reception of the Blessed and Cursed Humans, after their Resurrections; but, such as these Worlds we are of, only freely Regular, or Irregular."[46] Is this a sophistical way of disowning a theology by definition dubious from a Christian point of view?[47] We have seen that Cavendish repudiated any charge that she might be an atheist. In most of her natural philosophic writings, she professed her desire to avoid theological subjects, even when they actually contained a great deal of theology. But here she changes the focus of her discussion from the afterlife to this life. Thus, she takes as her point of orientation this world, which she calls a "Purgatory" world because it is "partly Irregular, and partly Regular." Besides this world, there are many regular worlds where the rational and sensitive parts live in harmony. Like Heaven, the regular world is a perceptual paradise:

> But, not to be tedious; it was my Mind's opinion, That all the Parts of the Happy World, being Regular, they could not obstruct each other's Designs or Actions; which might be a cause, that both the Sensitive and Rational Parts may not only make their Societies more curious, and their Perceptions more perfect; but their Perceptions more subtile: for, all the actions of that World being Regular, must needs be exact and perfect; in so much, that every Creature is a perfect Object to each other; and so every Creature must have, in some sort, a perfect Knowledg of each other.[48]

The result is that "all the particular Human Societies, (which are particular Human Creatures) live as if they were but one Soul, and Body; that is, as if they were but one Part, or Particular Creature."[49] All of the error—and sin—that Cavendish had argued was the consequence of the limited knowledge of Nature is avoided in the regular worlds. This description clarifies the one major change

Cavendish made in the natural philosophic section of *Grounds of Natural Philosophy*. She explains, "I was of an opinion, That Nature, because Infinite, could not know her Self; because Infinite hath no limit. Also, That Nature could not have an Absolute Power over her own Parts, because she had Infinite Parts; and, that the Infiniteness did hinder the Absoluteness: But since I have consider'd, That the Infinite Parts must of necessity be Self-knowing; and that those Infinite Self-knowing Parts are united in one Infinite Body, by which Nature must have both an United Knowledg, and a United Power."[50]

In the regular world, Nature and her parts are subsumed into one unit: a kind of Parmenidean whole, allowing for distinctions between particulars without sacrificing their basic unity. Even infinity does not limit the possibility of Nature knowing herself and being unable to control her parts. Indeed, all of the parts share in the omniscience and omnipotence of nature. The result is that the inhabitants of the regular worlds are happy: "But surely, all Human Creatures of that World, are so pleasant and delightful to each other, as to cause a general Happiness."[51]

In the regular world, since there are "no Irregularities, all Creatures must needs be Excellent, and most Perfect, according to their Kind and Sort." Their lives are full of "pleasures, and Pleasant Pastimes." Among other irregularities the citizens of this happy world avoid, "there are no Plots or Intrigues, neither in their State, nor upon their Stage."[52] In some other worlds, in other words, there are no plays of the kind Cavendish was publishing at the same time as *Grounds of Natural Philosophy*. Lady Happy would not need to create a Convent of Pleasure in the regular world because all possibility of pleasure was already present—although what these particular pleasures might be, Cavendish explains, "not any Creature can express, unless they were of that World, or Heaven."[53]

It is even more difficult "to conceive those Irregularities that are in the Irregular World," but Cavendish tries to do it. The irregular world is like the Blazing World, but distorted and turned upside down by something like Cartesian vortices. It is worth repeating most of the entire second chapter of part 3 of the appendix of *Grounds of Natural Philosophy* because it brings together so many of the themes in Cavendish's work. She writes,

> According to the Actions of Nature, all Creatures are produced by the Associations of Parts, into particular Societies, which we name, Particular Creatures: but, the Productions of the Parts of the Irregular World, are so Irregular, that all Creatures of that World are Monstrous; neither can there be any orderly or distinct kinds and sorts; by reason that Order and Distinction, are Regularities.

Wherefore, every particular Creature of that World, hath a monstrous and different Form; insomuch, that all the several Particulars are affrighted at the Perception of each other; yet, being parts of Nature, they must associate; but their Associations are after a confused and perturbed manner, much after the manner of Whirlwinds, or Aetherial Globes, wherein can neither be Order, nor Method.[54]

The creatures of the irregular world are monstrous, although composed of the same kind of matter characterizing Cavendish's vitalistic materialism. They share the same kind of perceptions as their more blessed counterparts in the regular world, but their perceptions are so confused and perturbed that the creatures terrify each other. They cannot escape the subjective reality of each other's disorder, and so they suffer.

What they see is indeed frightening, but not unfamiliar if they had managed to read *Blazing World*. "By reason of the Irregularities, they are strangely mixt and disordered," and "being of confused Shapes or Forms, none of those Animal Creatures can be said to be of such, or such a sort."[55] These are the creatures the Fool had dreamed of in "The Presence." These are the beast-men of the Blazing World in a perpetual state of agony because of the hybrid nature of their forms. Forgetting for a moment that she is describing the irregular world and not Hell, Cavendish writes they "may appear as several Devils to each other."[56] Here, then, is another analogical meaning for beast-men that reveals the extent to which hermaphroditical figures still haunted—and horrified— Cavendish's imagination.

Not even death can release the mixed-up monsters from their irregular world. Even if their parts dissolve, "those dispersed Parts cannot joyn to any other Society, but what is as bad as the former."[57] Thus, the parts persist in one form or another. The same is true of the inhabitants of the regular world, who suffer death but are happy in the knowledge that their parts will be reconstituted into happy creatures, "so that, though the particular Human Creatures did dissolve from being Humans; yet, their Parts could not be Unhappy, when they did unite into other Kinds, and Sorts."[58]

But such universal dissolution and recomposition, whether regular or irregular, does not comfort the parts of Margaret Cavendish's mind: "They were sad, to think their kind Society should dissolve, and that their Parts should be dispersed and united to other Societies, which might not be so friendly as they were."[59] Cavendish's mind is here contemplating the mortality implicit in her own material philosophy. The possibility of any particular creature being recomposed from its dispersed parts, which are now integrated into other bodies, is extremely unlikely. Earlier in *Grounds of Natural Philosophy*, Cavendish had considered the likelihood of bodily recomposition and concluded, "It may not be impossible: but yet, It is very improbable, that such numerous sorts of Motions, after so general an Alteration, should so generally agree in an unnatural action."[60] The unnatural action she means is the reassembling of the body by its active parts.

Cavendish had discussed the possibility of bodily regeneration once before. In *Natures Pictures*, a traveler descends to the center of the earth, where he discovers that all dead creatures—human, animal, vegetable, and mineral—are combined in a sea of blood, thus losing their bodily integrity and possibility of regeneration.[61] Cavendish returned to this question in *Grounds of Natural Philosophy*. A book that began with the plea to the universities of Europe to allow her philosophy, if they disliked it, "to lye still in the soft and Easie Bed of Oblivion" ends with a consideration of whether material being could be resuscitated by what she calls "restoring beds" or "wombs." In a self-reflective conversation between the major and minor parts of her mind, the conclusion is reached "that if the Roots, Seeds, or Springs of a Society, or Creature, were not dissolved or dispersed, those Creatures might be restored to their former condition of Life, if they were put, or received, into the Restoring Beds."[62]

These restoring beds would be made of a kind of flesh, unlike the similarly transformative philosopher's stone, whose existence Cavendish denied.[63] The restoring beds restore what once existed rather than metamorphosing one kind of matter into another, as the "chymists" claimed.[64] In fact, Cavendish postulated, there might be one primary restoring bed on a small island in the very center of the sea. This being is

a Creature, like (in outward Form) to a great and high Rock: Not that this Rock was Stone; but, it was such a nature, (by the natural Compositions of Parts) that it was compounded of Parts of all the principal Kinds and Sorts of Creatures of this World, viz. Of Elemental, Animal, Mineral, and Vegetable kinds: and being of such a nature, did produce, out of it self, all kinds and sorts of Restoring-Beds; whereof, some sorts were so loose, that they only Hung by Strings, or Nerves: others stuck close. Some were produced at the top, or upper parts: others were produced out of the middle parts; and some were produced from the lower parts.[65]

If one were lucky enough, or quick enough, to get the remains of any creature to the restoring beds, its flesh would grow again, and it would be restored to its former being. The trick is to keep all the parts together, which is why, explains

Cavendish's mind, disembowelment is particularly sad, and "it was a greater cruelty to murder a dead man, and to rob him of his Interior Parts; than to murder a living man, and yet suffer his whole Body to lye peaceably in the Urn, or Grave."[66] Dry bones, in the biblical formulation, would indeed arise again, by means of the restoring beds.

Cavendish is very emphatic in distinguishing the natural action of the restoring beds from the artificial works of the experimental philosophers, who "wast their Time and Estates, with Fire and Furnace, cruelly torturing the Productions of Nature, to make their Experiments."[67] And although the restoring beds are attached to some kind of fleshly rock, they are not like the "Philosophers-Stone, which the Chymists believe to be some Deity, that can restore all Sorts and Kinds."[68] Men and their instruments cannot produce and restore life, but Nature's womb—her "Breeding-Beds," as Cavendish also calls the restoring beds— can naturally recompose the parts of matter. Hanging on to the rock in the middle point of the center of some undiscovered sea, the restoring beds, attached by umbilical cords, labor to produce life: "That the Animal Restoring Bed, was of such a Nature or Property, that it could dilate and contract, as it had occasion; in so much, that it could contract to the compass of the smallest, or to the extend to the magnitude of the largest Animal . . . and that when an Animal Creature was put into the Restoring-Bed; it would immediately inclose the Animal: and when it had caused a perfect Restoration, the Restoring Bed would open it self, and deliver it to its own Liberty."[69]

What are we to make of this materialist fancy? It seems to anticipate a major theme of future science fiction: Frankenstein's monster's vivification from the remains of corpses and the cannibalistic plant of *The Little Shop of Horrors* are anticipated in the meditations of the first woman natural philosopher. Cavendish's *Blazing World* and her other imagined worlds were works of science fiction. This last speculation seems more akin to horror. The image of a restoring bed, a kind of grotesque womb, shows Cavendish reimagining science and religion in a way informed by the gendered experience of birth. Cavendish saw her work as her child, and she refashioned it constantly throughout her writing career. In a certain sense, she was a restoring bed herself, constantly taking the fragments of her former work and breathing new life into them. Her huge volume of work was in a sense as monstrous as the product of a labor that lasted years. To her mind, it deserved to be wondered at and admired.

Cavendish combined elements of natural philosophy and female experience to make sense of the world, although even at the end of *Grounds of Natural Philosophy* she was not optimistic about the fate of her compositions/productions. In the conclusion to the discussion about restoring beds, the Skeptical Parts of her brain emerge from the "glandula" or kernel, where they have been residing, to criticize the other parts of Cavendish's brain. Equating the speculations of the Major Parts and Minor Parts of the Brain with the esoteric—and unprovable—notions of "chymists," the Skeptical Parts advise them to suppress their speculations. Otherwise, when men unsuccessfully try to find the restoring beds, they will be thrown into despair and waste their estates, "for, your Books send men to Sea, a much Cooler Element than Fire; but, more Dangerous than Chymical Fire, unless Chymical Fire be Hell-Fire."[70] While the rest of the parts of the mind give the Skeptical Parts the boot for causing unrest, Cavendish's own self-doubt is clear. Can fancy be attached to philosophy—can a woman do natural philosophy without being a monster or a fool? Or do fools and monsters have a particular kind of intuition not shared by those who investigate nature more soberly?

Cavendish's eschatology, whether focused on the material regeneration of parts by a restoring bed or womb, or on the resurrection of a material soul in a material body, or on the necessity of other worlds to house sinners and the saved, reflected the mind of a woman liberated by the philosophic and theological changes of the sixteenth and seventeenth centuries. Her own insistence on the agency of the mind in producing not only knowledge but perhaps the very objects of its contemplation allows us to attempt to see how she saw and created her worlds. Cavendish provides a view of how the new science, the new world, and the new religions ignited the understanding and the imagination of those, either female or uneducated, who did not attend universities or meetings of scientific societies.

# Does Cavendish Matter?

The cultural historian Robert Darnton writes in the introduction to his classic study of early modern culture, "When we cannot get a proverb, or a joke, or a ritual or a poem, we know we are on to something. When picking at a document where it is most opaque, we may be able to unravel an alien form of meaning. The thread might even lead into a strange and wonderful world view."[1]

If we read Margaret Cavendish as a *stupor mundi*, a wonder of the world to whom we seek to attach meaning, the complexities of admitting women (or at least a woman) into the history of the scientific revolution become apparent. Viewed as deviant in her own time and remembered as "eccentric" or "mad" until very recently, Cavendish makes a modern effort to understand her an exercise in either hermeneutics or apology.

Margaret Cavendish understood that she was not like other women or like other natural philosophers; she felt discomfited by her place in courtly and scientific societies. She was neither here nor there; her gender precluded acceptance within the emerging community of naturalists and experimentalists, and her shyness undermined participation in the courtly and witty society of Restoration England. Her acute sense of self propelled her fascination with things that were more than themselves: hybrids, monsters, other worlds.

Cavendish responded to her sense of alienation and her own intoxication with new ideas by making her own scientific revolution. The more science became contained and controlled under the aegis of the Royal Society, the more extravagant and speculative her own natural philosophy became. The more other investigators of nature limited their conclusions to what they could see, the more Cavendish credited the primacy of conception and reason. While other investigators of nature adopted the scientific report as the only legitimate way of conveying information, Cavendish embraced imagination and romance. Her natural philosophy, consequently, although incorporating many of the same elements and assumptions as that of her contemporaries, is itself different and unique.

Perhaps the best way to understand Margaret Cavendish is to see her as one of the hybrids who both enthralled and repelled her. The comparison applies on many levels. Cavendish was aware of the incongruity of her own actions and writing. She was sensitive to the accusation that women were essentially bestial or irrational. She knew that she was a cultural hermaphrodite, attempting to transcend her sex by writing instead of cooking or spinning. Her self-knowledge and constant self-reflection were tinged with self-doubt but also a belief in her own abilities. She knew how her own times would judge her, and she was not impervious to the voices of condemnation she accurately anticipated. Liberated to write by the extraordinary changes she experienced in the political and philosophic world of the mid-seventeenth century, and enabled by a supportive husband, Cavendish nevertheless was conscious of the audacity of her effort. Whatever she did was not easy, but it was reasoned. She was neither silly nor mad.

Cavendish defended herself, and her right to be a natural philosopher, by challenging the honor and position of the Royal Society during her spectacular visit to its college and in transmuting the experimentalists and natural philosophers into beast-men subjects of the Empress in *Blazing World*. Such a concerted, wide-ranging, and clever attack on many of the perceptions and pretensions of the new scientists, in many different styles of writing and self-representation, testifies to the ultimate sanity of Margaret Cavendish. But did she mean for her wit to serve a feminist agenda?

Cavendish's feminism erupts from her own historical context. Since notions of sex, nature, and "feminism" itself have meanings only in a culturally and temporally specific way, it is imperative to see Cavendish from the perspective of her own time. Cavendish certainly was not a feminist if feminism is taken to mean the empowering of all women. She had low expectations for the members of her sex, even if their insufficiencies were the result of poor education and male tyranny, as well as self-doubt about her own abilities. She did not privilege female reason; nor did her theory of matter contain an internalized female principle. Matter as matter was not sexed. Nevertheless, Margaret Cavendish's natural philosophy was gendered.

Cavendish was certainly aware of the potential ambiguity of sexuality and gender: hence her fascination with hybrid forms. She capitalized on her own unique self-gendered representations to gain the fame she so insistently wanted.

Gender enabled Cavendish to write the way she wanted to write. The seventeenth-century notion of proper womanhood as domestic and dumb became the foil for Cavendish's self-fashioning. The repetitive, conflicted, and urgent tone of her works, as she tried to make herself understood and admired in a new world of intellectual activity barred to women, demonstrates that gender categories could be liberating as well as constricting. By defining herself by what she was not—a "chymist," an experimental philosopher, a silly woman—Cavendish could become, at least in her own mind, a natural philosopher and a scientific revolutionary in her own right.

The search for certainty in the sixteenth and seventeenth centuries inspired many thinkers to search for new answers to replace those discredited by discovery and conflict. Knowledge itself became problematic. Michel de Montaigne asked how anyone could know anything, and the protagonists in the *querelle des femmes* wondered particularly about whether a woman could know anything. "Who knows?" Cavendish repeatedly asks in her earlier works, but then she uses epistemological uncertainty as a wedge to enter the all-male world of natural philosophy. Thus Cavendish, in her earliest works, *Poems, and Fancies* (1653) and *Philosophicall Fancies* (1653), used her fancy to imagine a world of tiny living parts—atoms—that were essentially the same as fairies. She envisioned a world where Nature commanded and matter and motion followed. She endowed matter with vitality and feeling. And she claimed parity with other philosophers of nature, since her ideas were just as probable as theirs. Cavendish grounded her own thought on what she called "sense and reason," or the knowledge she gained and produced through her own body and mind. Excluded from universities and scientific societies, Cavendish issued herself a passport to breach borders of increasingly masculine places. Essentially, as the science around her became more empirical, her natural philosophy became more speculative. She felt no need to contain her ideas within the parameters of either experiment or method.

The belief in the capacities of unmediated sense and reason to observe and understand nature increasingly became Cavendish's warrant for writing natural philosophy. Obviating the need for formal education, sense and reason became epistemological tools for an exploration of any number of ideas about the nature and possibilities of matter. In fact, in the works Cavendish published in 1655 and 1656, *Philosophical and Physical Opinions* and *Natures Pictures*, reason and sense themselves are expressed as two different forms of matter: innate matter, composed of rational and sensitive parts, and an inanimate kind of matter, which together make up the totality of material being. Cavendish recognized that in proposing the agency and self-motion of matter she implicitly challenged traditional religion and authority, a challenge she tried to excuse. Religion and philosophy, she argued, should be separated—but this was a divorce that she, like many of her peers, was never able to achieve. Materialism haunts her philosophy, and although she claimed that there was a difference between the immortal soul and the corporeal soul, her unwillingness to accept the notion of any incorporeal form made her protestations sound hollow. Cavendish undoubtedly believed in God and was a faithful, if not passionate, member of the Anglican Church, but her adherence to the idea of naturally eternal and infinite matter trumped any sympathy with spirituality in her thought.

Cavendish's willingness both to utilize and to challenge many of the social, intellectual, and religious norms of her own society was rational in her own terms, even if irrational in terms of seventeenth-century expectations. Since she was composed of rational and sensitive matter (with a little inanimate matter thrown in), she was acting rationally in composing natural philosophy. The nature of rational matter is to think. Cavendish could reply to Descartes: I am, therefore I think. But she suggested in *Natures Pictures* that thought itself, being material, is full of error, and perhaps even sin. The problem of understanding matter and its aberrations, therefore, becomes the problem of understanding the nature of nature.

What is nature? We might say our current effort to historicize the meaning of nature was anticipated by Cavendish's attempt to problematize the meaning of the term. For Cavendish, nature could be either the internalized vital force of matter or an externalized anthropomorphic ruler of matter. Cavendish had many definitions of nature: Nature, personified as a woman, is a creator, but nature is also the matter constituting the universe. The gendered Nature is the servant of God, or a housewife, or a benevolent matron. This Nature is the powerful arbiter and director of the world, while the world itself is composed of the material parts of nature. Both generating and self-generated, nature is all-encompassing, self-moving, and self-conscious, whether as an actual being or the totality of being.

Some of the confusion in understanding Cavendish's philosophy is because of how she played with these two representations of nature. In the 1663 revision of *Philosophical and Physical Opinions*, she developed a theory of holistic matter that retained the notion of a hierarchy of matter—rational, sensitive, and inanimate—but a hierarchy so integrated and entwined that any diversity is subsumed into unity. In 1663, with the experience of the civil war vivid in her mind, she allowed that there might be some form of conflict inherent in matter because each part lacks knowledge of the infinite whole and Nature does not have an infinite knowledge of her parts. But each part, when organized into different kinds, does possess the rationality and knowledge of its particular kind of being. The idea of the equality of kinds is presented as response and repudiation of human arrogance and presumption. Species-specific rationality, which is a natural product of the rationality of matter, undermines the idea of the mechanistic universe. Nature can neither be forced by impact, as Hobbes would have it, nor separated into extended and unextended substances, as Descartes insisted. There is no spirit, immaterial or preternatural, that animates matter and gives it understanding. Everything in the universe is vitally alive and understands everything else, at least as far as its kind and capacity allow. The stone knows like a stone, the animal like an animal, the vegetable like a vegetable, and the woman like a woman—or perhaps, like a human being.

Did Cavendish believe that there was a particular woman's way of knowing? Early in her career she suggested that women are particularly adept at imaginative knowledge, and later she included imagination as a form of rationality. Nevertheless, the mechanisms of perception are not particular to any one gender or species: all creatures learn by the sensitive organs patterning exterior objects, which then are understood in the mind, which generates its own material thoughts. These thoughts are autonomous and in a sense constitutive of the reality the mind understands. The difference between kinds comes down to a difference in the quantity of rational matter they possess. Thus, Cavendish was able to maintain an egalitarian epistemology allowing for distinction among different beings but never surrendering to a hierarchy of rationality. Humans may have more rational matter, but they are no better than animals, vegetables, or even stones.

A homogeneous matter underlies a universal kind of equality. Cavendish's later works, in particular those written after she had studied works by the major natural philosophers of her time, can be seen as a defense of all kinds of beings against those who would presume to command or even understand them. Ringingly, in *Philosophical Letters* (1664), the duchess defends Nature, women, and witches from the manipulation of "chymists" and experimental philosophers and the enervating presence of immaterial principles. As Cavendish's work developed and she realized that she would never be allowed to join the community of natural philosophers, she began to emphasize the imperial power of a personified Nature impervious to manipulation and to populate her fictions with commanding women. Error is no longer an integral aspect of Nature, as Nature becomes more than the sum of her parts. Nature produces all created being, either immediately or by providing the tools and material for objects produced by art. Her own creations are infinitely better than these artificial products of "chymists" or experimenters, even when they work with what she has made.

In *Observations* and *Blazing World*, published in one volume in 1666 and re-printed in 1668, Cavendish defends what we might call the life of Nature against those who might seek to derogate her power. She equates experimenters with boys, who presume against the natural productions of the creator, Nature. She con-demns their instruments, and the knowledge they produce, as artificial mon-strosities that ultimately reveal nothing but the futility of trying to penetrate na-ture. In the romance *Blazing World*, foolish philosophers become akin to the objects they investigate. Beast-men weigh air and look at the stars. They are all, except for the worm-men, deluded and useless. The threat to the state from their contentiousness ceases only when the Empress of the Blazing World uses their inborn abilities in her efforts to conquer other worlds. Just as Nature rules matter, the Empress reigns over natural philosophy and faction. As Cavendish describes her ambitions at the end of *Blazing World*, she desires to be not only "Emperess, but Authoress of a whole World," and she invites her readers to be her subjects: "They may imagine themselves such, and they are such; I mean, in their Minds, Fancies or Imaginations."[2]

But Cavendish's natural philosophy was not simply a response to the new sci-ence. She took what she had heard and read and transformed it into her own cre-ation. Her vision was animated by a cultural imagination, which was unrestrained by experiment, logic, or institutional conformity. In a sense, Cavendish was a re-storing bed, like the one in *Grounds of Natural Philosophy* (1668), securing and re-storing the vitality of nature. Margaret Cavendish's natural philosophy could have been written only by a seventeenth-century woman—indeed, only by one particu-lar seventeenth-century woman. No man of the time could have breached the categories she did in her work and her life. Her Nature is self-moving, self-conscious, and active, as was Cavendish herself. She could say, with an entire philosophic system to back her up: I create, I challenge, I produce, I transform.

Cavendish demonstrates how natural philosophy found an audience that was eager to explore the ramifications of the new science but whose response was not necessarily what the practitioners might have wanted. She may have been more successful than anyone has realized. At the beginning of Thomas Sprat's *History of the Royal Society* (1667), the poet Abraham Cowley sang,

> Mischief and tru Dishonour fall on those
> Who would to laughter or to scorn expose
> So Virtuous and so Noble a Design,
> So Human for its Use, for Knowledge so Divine.[3]

And in 1694, twenty years after Cavendish died, the English scholar and linguist William Wotton reflected on the continued difficulties the Royal Society had in being taken seriously: "For nothing wounds so much as a jest; and when men do once become ridiculous, their labours will be slighted and they will find new imitators. How far this may deaden the industries of the next age is not easie to tell."[4]

Cavendish belongs to a tradition of critics of modern science, a pattern extending from Ben Jonson to Thomas Shadwell—both clients of her husband's—and stretching forward to Jonathan Swift. She joined Cyrano de Bergerac and Bernard de Fontenelle in contemplating the implications of other worlds and anticipated modern writers of science fiction from Mary Shelley to Ursula Le Guin.[5] In these writers, the definition of what is human becomes fluid: instead of introducing order and method, unbridled science problematizes human knowledge and morality, creating monsters, not facts.

Cavendish had a sense of humor, something her opponents lacked and feared. Her reaction to the new philosophy showed that she had taken care to understand its claims and even to share its basic ontology of matter in motion in developing her own material philosophy. But satirists always have a serious purpose when they laugh at the pretensions of the powerful, or those who want to be powerful. Cavendish spoke for those objectified and excluded by the new science, and like many satirists she understood the implications of new ideas.

But Cavendish not only criticized mechanical and experimental philosophy; she developed her own explanation of nature. To some extent, her philosophy was anachronistic. While her contemporaries eschewed hypothesizing beyond the observed facts, Cavendish embraced speculation, which combined reason and fancy. Her version of vitalistic materialism could comfortably fit into the sixteenth-century paradigm of Giordano Bruno, Paracelsus, or even Johannes Kepler. By the late seventeenth century, however, system making was no longer in favor. There was no place for grand schemes in science, just as there was no space for a woman in the meetings of the Royal Society.

Ultimately, Cavendish was a conservative, even as she adapted many of the newest ideas of her time and produced a most unconventional natural philosophy. Like many revolutionaries seeking a new world and a golden age, Cavendish belonged to a past her own work showed was impossible to resurrect. The restoring bed of the Restoration did not bring back her husband's status or generate the respect she thought she was owed. The values and ideas she wanted to preserve were the traditional ones of class and hierarchy. Such a view of society complemented Cavendish's natural philosophy. Each part has its place, but all parts are ruled by a woman whose flesh they constitute. Such rule, after all, is natural. The playwright Thomas Sprat understood Margaret Cavendish's claim to glory, and in his eulogy written after her death in 1673, he emphasized both her importance and her singularity:

> Philosophers must wander in the dark;
> Now they of Truth can find no certain mark;
> Since She their surest Guide is gone away,
> They cannot chuse but miserably stray.
> All did depend on Her, but She on none,
> For her Philosophy was all her own.
> She never did to the poor Refuge fly
> Of Occult Quality or Sympathy.
> She could a Reason for each Cause present,
> Not trusting wholly to Experiment,
> No Principles from others she purloyn'd,
> But wisely Practice she with Speculation joyn'd.[6]

Historians of science no longer ignore Cavendish's ironic analysis of the scientific revolution, which ultimately turns it upside down and even feminizes the activities of male experimenters. *Observations upon Experimental Philosophy* has just been added to the Cambridge Texts in the History of Philosophy series. Cavendish even appears in Western civilization texts, a development that would have delighted and amused her. Her inclusion emphasizes that the scientific revolution was not uncontested and that it was perhaps a woman who was best able to challenge the pretensions and power of the new science.

# Notes

BW            *Description of a New World, Called the Blazing World* (London, 1666)

BW 1668      *Description of a New World, Called the Blazing World*, stand-alone ed. (London, 1668)

GNP          *The Grounds of Natural Philosophy* (London, 1668)

Life 1916     *The Life of the (1st) Duke of Newcastle and Other Writings by Margaret Duchess* (London: J. M. Dent & Sons, 1916)

NP           *Natures Pictures* (London, 1656)

OEP         *Observations upon Experimental Philosophy* (London, 1666)

P & F        *Poems, and Fancies* (London, 1653)

PF            *Philosophicall Fancies* (London, 1653)

PL            *Philosophical Letters* (London: 1664)

PPO 1655    *Philosophical and Physical Opinions* (London, 1655)

PPO 1663    *Philosophical and Physical Opinions*, 2nd ed. (London, 1663)

SL            *Sociable Letters* (London, 1664)

TR           *A True Relation of my Birth, Breeding, and Life* (London, 1656), in *Paper Bodies: A Margaret Cavendish Reader*, ed. Sylvia Bowerbank and Sara Mendelson (Peterborough, Ontario: Broadview Press, 2000)

WO          *The Worlds Olio* (London, 1655)

*Note:* Unless noted otherwise, all citations of these and other seventeenth-century works are to the published originals.

## INTRODUCTION: GENDER, NATURE, AND NATURAL PHILOSOPHY

1. *BW*, 56.

2. *BW*, 28.

3. *BW*, 122–23. Lawrence Principe argues that *chymistry* is the most accurate term for indicating "the sum total of alchemical/chemical topics as understood in the seventeenth century." (Lawrence M. Principe, *The Aspiring Adept: Robert Boyle and His*

*Alchemical Quest* [Princeton: Princeton University Press, 1998], 9.) *Chymistry* is the term Cavendish uses when discussing alchemical activities.

4. The clearest statement of this view is Theodore K. Rabb, *The Struggle for Stability in Early Modern Europe* (Oxford: Oxford University Press, 1976).

5. Henrietta Maria (1609–69) was the wife of Charles I (1600–1649), who fled with him to Paris after the first defeats experienced by royalist forces against the parliamentary rebels during the opening years of the English Civil War.

6. On the meaning of "gender" as a category of interpretation in the history of science, see Londa Schiebinger, *Has Feminism Changed Science?* (Cambridge, MA: Harvard University Press, 1999), 1–18.

7. Merry E. Wiesner, *Women and Gender in Early Modern England* (Cambridge: Cambridge University Press, 2000), 20–26.

8. "To the Reader," in *PPO* 1655, n.p.

9. Cavendish recounted and rationalized his military exploits in The *Life of the thrice noble, High and Puissant Price William Cavendishe, Duke, Marquess and Earl of Newcastle* (London, 1667). I am using the Everyman's Library edition of this text, abbreviated as "*Life* 1916" in the notes. Anna Battigelli discusses the role exile played in provoking Cavendish's ideas in *Margaret Cavendish and the Exiles of the Mind* (Lexington: University of Kentucky Press, 1998).

10. On Newcastle, see Geoffrey Trease, *Portrait of a Cavalier: William Cavendish, First Duke of Newcastle* (New York: Taplinger, 1979); Lucy Worsley, *Cavalier: A Tale of Chivalry, Passion, and Great Houses* (New York: Bloomsbury, 2007); and Lisa T. Sarasohn, "Thomas Hobbes and the Duke of Newcastle: A Study in the Mutuality of Patronage before the Establishment of the Royal Society," *Isis* 90 (1999): 715–37. His two books on manage are *La Méthode de invention nouvelle de dresser les chevaux* (Anvers, 1657–58) and *A New Method and Extraordinary Invention to Dress Horses, and Work Them According to Nature; as Also to Perfect Nature by the Subtlety of Art; Which Was Never Found Out But by the Thrice Noble, High and Puissant Prince, William Cavendish, Duke of Newcastle* (London, 1667). The treatise on swordsmanship was not published but can be found in a manuscript copy in the British Library (Harley MS 4206, fols. 2–3, 8–9).

11. Charles Cavendish's role in the scientific community is explored in the introduction to Noel Malcolm and Jacqueline Stedall, eds., *John Pell (1611–1685) and His Correspondence with Sir Charles Cavendish: The Mental World of an Early Modern Mathematician* (Oxford: Oxford University Press, 2005). Also see Jean Jacquot, "Sir Charles Cavendish and His Learned Friends," *Annals of Science* 8 (1952): 13–27; Helen Hervey, "Hobbes and Descartes in the Light of Some Unpublished Letters of the Correspondence between Sir Charles Cavendish and John Pell," *Osiris* 10 (1952): 69–90; and Mordechai Feingold, "Descartes and the English: The Cavendish Brothers," in *La Biografia Intellettuale de René Descartes attraverso La Correspondance* (Naples: Vivarium, 1999).

12. "To Natural Philosophers," in *P & F*, n.p.

13. "An Epistle to justifie the Lady Newcastle," in *PPO* 1655, n.p.

14. On women and the scientific revolution, see Londa Schiebinger, *The Mind Has No Sex? Women in the Origin of Modern Science* (Cambridge, MA: Harvard University Press, 1989).

15. Rabb, *Struggle for Stability*, 107–15.

16. On this historiographical reinterpretation, see Betty Jo Dobbs, "Newton as Final Cause and First Mover," *Isis* 85 (1994): 633–43, and Margaret J. Osler, ed., *Rethinking the Scientific Revolution* (Cambridge: Cambridge University Press, 2000), particularly the chapter by Osler, "The Canonical Imperative: Rethinking the Scientific Revolution," 3–22.

17. Most of this biographical information comes from *Life* 1916 and Cavendish's own autobiography, *A True Relation of my Birth, Breeding, and Life* (1656) in *Paper Bodies: A Margaret Cavendish Reader*, ed. Sylvia Bowerbank and Sara Mendelson (Peterborough, Ontario: Broadview Press, 2000), which was originally published in the same volume as *Natures Pictures* (London, 1656). This modern edition of *True Relation* is abbreviated as "*TR*" in the notes. There are two modern biographies of Margaret Cavendish: Battigelli, *Margaret Cavendish and the Exiles of the Mind*, and Katie Whitaker, *Mad Madge: The Extraordinary Life of Margaret Cavendish, Duchess of Newcastle, the First Woman to Live by Her Pen* (New York: Basic Books, 2002).

18. The most recent discussion of Cavendish's visit to the Royal Society is Peter Dear, "Understanding Margaret Cavendish and the Royal Society," in *Science, Literature and Rhetoric in Early Modern England*, ed. Juliet Cummins and David Burchell (Hampshire, England: Ashgate, 2007), 125–42.

19. Cavendish, "To the Reader," in *PL*, n.p.

20. *PL*, 2.

21. Cavendish, "To the Reader," in *OEP*, n.p.

22. Cavendish, preface to *PPO* 1663, n.p.

23. Cavendish, "To the Two Universities," in *PPO* 1655, n.p.

24. Cavendish, "The Preface to the Reader," in *WO*, n.p.

25. *SL*, 31.

26. *TR*, 43.

27. On Aristotelian and other medical views of women, see Ian Maclean, *The Renaissance Notion of Women: A Study in the Fortunes of Scholasticism and Medical Science in European Intellectual Life* (Cambridge: Cambridge University Press, 1980), and Schiebinger, *Mind Has No Sex?* 160–89.

28. Cavendish, "The Preface to the Reader," in *WO*, n.p.

29. Here Cavendish clearly reflected the debate about women that preoccupied the late sixteenth and seventeenth centuries, when women did indeed rule. See Katherine Usher Henderson and Barbara F. McManus, *Half Humankind: Contexts and Texts of the Controversy about Women in England, 1540–1640* (Urbana: University of Illinois Press, 1985).

30. Cavendish, "The Preface to the Reader," in *PPO* 1655, n.p.

31. Cavendish, "Youths Glory, and Deaths Banquet," in *Plays* (London, 1662), 134.

32. Ibid., 140.

33. Lorraine Daston, "The Nature of Nature in Early Modern Europe," *Configurations* 6, no. 2 (1998): 155.

34. Robert Boyle, *A Free Inquiry into the Vulgarly Received Notion of Nature* (1664), in *Robert Boyle on Natural Philosophy*, ed. Marie Boas Hall (Westport, CT: Greenwood Press, 1965), 150–53.

35. Robert Boyle, *The Excellency and Grounds of the Corpuscular or Mechanical Philosophy* (1674), in *Robert Boyle on Natural Philosophy*, 191, 196.

36. Cavendish, "Further Observations upon Experimental Philosophy," in *OEP*, 25.

37. Ibid., 6–7, 26.

38. Shiebinger, *Has Feminism Changed Science?* 4–5.

39. On Cavendish's use of "nature" as strategic ploy, see Sylvia Bowerbank, *Speaking for Nature: Women and Ecologies of Early Modern England* (Baltimore: Johns Hopkins University Press, 2004), 5, 53. On the metaphorical meaning of nature in the Middle Ages and early modern Europe, see Katharine Park, "Nature in Person: Medieval and Renaissance Allegories and Emblems," in *The Moral Authority of Nature* (Chicago: University of Chicago Press, 2004), 50–73.

40. Cavendish, "Further Observations," 41.

41. Ibid., 41, 45–46.

42. Ibid., 12, and Cavendish, "Female Orations," from *Orations of Divers Sorts, Accommodated to Divers Places* (1662), in Bowerbank and Mendelson, *Paper Bodies*, 143–47.

43. Cavendish, "Further Observations," 72.

44. Ibid., 4, 8.

45. Ibid., 25.

46. For the cultural meanings of human and animal in early modern Europe, see Erica Fudge, Ruth Gilbert, and Susan Wiseman, *At the Borders of the Human: Beasts, Bodies and Natural Philosophy in the Early Modern Period* (Hampshire, England: Palgrave, 2002), particularly Erica Fudge, "Calling Creatures by Their True Names: Bacon, the New Science and the Beast in Man," 91–109, and Susan Wiseman, "Monstrous Perfectability: Ape-Human Transformations in Hobbes, Bulwar, Tyson," 215–38.

47. Cavendish, "Further Observations," 11.

48. Ibid., 16.

49. *PL*, 40–41.


1: A WONDERFUL NATURAL PHILOSOPHER

1. The notorious former Queen Christina of Sweden was touring the various capitals of Europe at this time and attracting huge crowds. See Samuel Pepys, *The Diary of Samuel Pepys*, 10 vols., ed. Robert Latham and William Matthews (Berkeley: University of California Press, 1974), 8:164n1.

2. Ibid., 163–64, 196.

3. There were many thinkers who continued to accept the idea of an organic and living nature, including members of the Royal Society (see Christoph Meinel,

"Early Seventeenth-Century Atomism: Theory, Epistemology, and the Insufficiency of Experiment," in *The Scientific Enterprise in Early Modern Europe: Readings from Isis*, ed. Peter Dear (Chicago: University of Chicago Press, 1997), 176–211, and B. J. T. Dobbs, "Newton's Alchemy and His Theory of Matter," in Dear, *Scientific Enterprise in Early Modern Europe*, 237–54. The argument that the new science was hostile to an organic worldview is articulated most forcefully by Carolyn Merchant, *The Death of Nature: Women, Ecology and the Death of Nature* (San Francisco: Harper & Row, 1980).

4. According to Lorraine Daston and Katharine Park, *Wonders and the Order of Nature, 1150–1750* (New York: Zone Books, 1998), "Monsters inspired repugnance because they violated the standards or regularity not only in nature, but also in society and the arts. A monstrous birth undermined the uniform laws God imposed upon nature; the 'monstrous regiment of women' rulers threatened the order of civil society; the intrusion of marvels into poems and plays destroyed literary verisimilitude" (212).

5. Joan Kelly, "Early Feminist Theory and the *Querelle-des-Femmes*, 1400–1789," *Signs: Journal of Women in Culture and Society* 8 (1982): 19.

6. Deborah Boyle, "Margaret Cavendish's Natural Philosophy," *Configurations* 12, no. 2 (2004): 195–227.

7. *PL*, 1.

8. On the skeptical tradition, see Richard H. Popkin, *The History of Scepticism from Erasmus to Descartes*, rev. ed. (New York: Harper & Row, 1964).

9. *P & F*, 36.

10. *P & F*, 162. On the connection between fancy and reason in Cavendish's natural philosophy, see Rosemary Kegl, "'The World I Have Made': Margaret Cavendish, Feminism, and the *Blazing World*," in *Feminist Readings of Early Modern Culture: Emerging Subjects*, ed. Valerie Taub, M. Lindsay Kaplin, and Dympha Callaghan (Cambridge: Cambridge University Press, 1996), 127.

11. Cavendish, "To the Reader," in *NP*, n.p.

12. Cavendish, "To the Reader," in *BW*, n.p.

13. Kate Lilley has expertly described the hermaphroditical qualities of Cavendish's *Blazing World*, which not only tells the story of a virago warrior queen and her hybridized subjects but "is already improbably and hermaphroditically coupled with a serious treatise on natural philosophy, *Observations Upon Experimental Philosophy*," in *The Blazing World and Other Writings*, ed. Kate Lilley (London: Pickering, 1992; repr., London: Penguin, 1994), xxiv.

14. *SL*, 4. Lord Denny directed this advice to Lady Mary Worth, who was the first English woman to publish an original prose work (The Sidney Homepage, www.english.cam.ac.uk/worth/biography.html, accessed January 18, 2009).

15. *P & F*, 121.

16. Francis Bacon, "Advancement of Learning," in *Advancement of Learning, Novum Organum, New Atlantis* (Chicago: Encylopedia Britannica, 1952), 33–38, 12–14, and Bacon, "Novum Organum," in *Advancement of Learning, Novum Organum, New Atlantis*, 123.

17. Walter Charleton to Margaret Cavendish, January 1, 1654, in *A Collection of Letters and Poems*, 146.

18. William Newcastle, "To the Lady Newcastle," in *PPO* 1655, n.p.

19. William Newcastle, "Epistle to justifie the Lady Newcastle," in *PPO* 1655, n.p.

20. Steven Shapin and Simon Schaffer, *Leviathan and the Air-Pump: Hobbes, Boyle and the Experimental Life* (Princeton: Princeton University Press, 1985).

21. "Newcastle's Advice to Charles II," in Thomas P. Slaughter, *Ideology and Politics on the Eve of the Restoration: Newcastle's Advice to Charles II* (Philadelphia: American Philosophical Society, 1984), 226–27. On Newcastle's advice, see James R. Jacob and Timothy Raylor, "Opera and Obedience: Thomas Hobbes and a Proposition for the Advancement of Moralitie by Sir William Davenant," *Seventeenth Century* 6 (1991): 215–25. Lucy Worsley explores Newcastle's acquaintance with popular culture in *Cavalier: A Tale of Chivalry, Passion, and Great Houses* (New York: Bloomsbury USA, 2007), 103, 114, 203–4.

22. *Life* 1916, 199.

23. Paula Findlen, "Between Carnival and Lent: The Scientific Revolution at the Margins of Culture," *Configurations* 6 (1998): 243–67, 253, and Findlen, "Jokes of Nature and Jokes of Knowledge: The Playfulness of Scientific Discourse in Early Modern Europe," *Renaissance Quarterly* 43 (1990): 292–331, 10.

24. Cavendish, "Further Observations upon Experimental Philosophy," in *OEP*, 13.

25. *Life* 1916, 136.

26. Cavendish, "To Natural Philosophers," in *P & F*, n.p.

27. Cavendish, "An Epistle to Mistris Toppe," in *P & F*, n.p.

28. *SL*, 204. According to Fitzmaurice in his introduction to *Sociable Letters*, Cavendish wrote this letter and several others describing popular culture when she lived in Antwerp (24).

29. *SL*, 205.

30. Natalie Z. Davis, "The Reasons of Misrule," in *Society and Culture in Early Modern France* (Stanford: Stanford University Press, 1965), 119.

31. Quoted in Katie Whitaker, *Mad Madge: The Extraordinary Life of Margaret Cavendish, Duchess of Newcastle, the First Woman to Live by Her Pen*, 156. Whitaker emphasizes that Cavendish's reputation for eccentricity was increased by her unusual sartorial style (155–56).

32. Davis, "Women on Top," in *Society and Culture in Early Modern France*, 130–31. For more on the topos of a world turned upside down, see Christopher Hill, *The World Turned Upside Down* (London: Penguin, 1972); Peter Burke, *Popular Culture in Early Modern Europe* (New York: Harper & Row, 1978), 185–204; and Jerome Nadelhaft, "The Englishwoman's Sexual Civil War: Feminist Attitudes towards Men, Women, and Marriage, 1650–1740," *Journal of the History of Ideas* 43 (1982): 55–66.

33. Cavendish's most recent biography is called *Mad Madge*, although its author, Katie Whitaker, shies away from endorsing this view of her subject. Sarah Hutton, "Anne Conway, Margaret Cavendish and Seventeenth-Century Scientific Thought," in *Women, Science and Medicine, 1500–1700: Mothers and Sisters of the Royal Society*, ed. Lynette Hunter and Sarah Hutton (Thrupp-Stroud, Gloucestershire, England:

Sutton, 1997), notes Cavendish's willingness to accept ridicule in return for recognition of her uniqueness: "In the prefaces to her books she [Cavendish] calls attention to the fact that women writing philosophy were considered an anomaly—'men in petticoats.' And she seems deliberately to have made a virtue of the eccentricity this implies in order to make a niche for her learned pursuits" (22).

34. Arthur O. Lovejoy, *The Great Chain of Being: A Study of the History of an Idea* (New York: Harper & Row, 1960).

35. Merchant, *Death of Nature*, 1–6.

36. Cavendish, "To the Reader," in *OEP*, n.p.

37. Cavendish, "Orations in the Field of Peace," in *Orations of Divers Sorts, Accommodated to Divers Places* (1662), in *Women's Political and Social Thought*, ed. Hilda L. Smith and Berenice A. Carroll (Bloomington: Indiana University Press, 2000), 79.

38. Cavendish, "To His Excellencie the Lord Marquis of Newcastle," in *PPO* 1663, n.p.

39. *SL*, 4.

40. *OEP*, 101–2.

41. Margaret L. King, "Book-Lined Cells: Women and Humanism in the Early Italian Renaissance," in *Beyond Their Sex: Learned Women in the European Past*, ed. Patricia H. Labalme (New York: New York University Press, 1980), 75. Although King is writing about the Italian Renaissance, her conclusion applies equally to the seventeenth century.

42. Daston and Park, *Wonders and the Order of Nature*, 14.

43. Samuel I. Mintz, "The Duchess of Newcastle's Visit to the Royal Society," *Journal of English and Germanic Philology* 51 (1952): 176. See also Marjorie Hope Nicolson, *Pepys' Diary and the New Science* (Charlottesville: University Press of Virginia, 1965), 110–11, and Whitaker, *Mad Madge*, 298–300. More nuanced treatments can be found in Sarah Hutton, "Anne Conway, Margaret Cavendish and Seventeenth-Century Scientific Thought," 223; Sylvia Bowerbank, *Speaking for Nature: Women and Ecologies of Early Modern England* (Baltimore: Johns Hopkins University Press, 2004), 64; and Peter Dear, "Understanding Margaret Cavendish and the Royal Society," in *Science, Literature and Rhetoric in Early Modern England*, ed. Juliet Cummins and David Burchell (Aldershot, England: Ashgate, 2007), 125–44. See the Essay on Sources for more detailed discussion of the historiography of the visit.

44. Rebecca D'Monté, "'Making a Spectacle': Margaret Cavendish and the Staging of Self," in *A Princely Brave Woman: Essays on Margaret Cavendish, Duchess of Newcastle*, ed. Stephen Clucas (Aldershot, Hampshire, England: Ashgate, 1988), 109.

45. *TR*, 57. On public display and collecting, see Paula Findlen, *Possessing Nature: Museums, Collecting, and Scientific Culture in Early Modern Italy* (Berkeley: University of California Press, 1994). Emma Rees, *Margaret Cavendish: Gender, Genre, Exile* (Manchester: Manchester University Press, 2003), notes that these kinds of displays were outlawed during the Commonwealth period (37). Hence, a public display was not only a cultural event but a political statement.

46. Burke, *Popular Culture in Early Modern Europe*, 43.

47. Ibid., 24–28.

48. *TR*, 43.

49. The question of what popular culture is and whether it can ever be accessed through mediating sources has preoccupied cultural historians for the past two decades. See Barry Reay, *Popular Cultures in England, 1550–1750* (London: Longman, 1998), and Tim Harris, ed., *Popular Culture in England, c. 1500–1850* (New York: St. Martin's Press, 1995), 1–27. Roger Chartier in *The Cultural Uses of Print in Early Modern Europe* (Princeton: Princeton University Press, 1987) argues that since the concept of popular (and elite) culture is so problematic, it is better to study how texts and practices are used or "appropriated" rather than the social class that produces them.

50. Quoted in Sara Heller Mendelson, *The Mental World of Stuart Women: Three Studies* (Amherst: University of Massachusetts Press, 1987), 46.

51. Cavendish, *Orations of Divers Sorts, Accommodated to Divers Places* (1662), in *Margaret Cavendish: Political Writings*, ed. Susan James (Cambridge: Cambridge University Press, 2003), 250.

52. On the role of hermaphroditical imagery in Cavendish, see Eve Keller, "Producing Petty Gods: Margaret Cavendish's Critique of Experimental Science," *ELH* 64 (1997): 447–71, and Nicole Pohl, "Of Mixt Natures: Questions of Genres in Margaret Cavendish's The Blazing World," in Clucas, *Princely Brave Woman*, 51–68.

53. Cavendish, "The Text to my Natural Sermon," in *PPO* 1655, n.p.

54. See, for example, Cavendish, "Love's Adventures," in *The Convent of Pleasures and Other Plays*, ed. Anne Shaver (Baltimore: Johns Hopkins University Press, 1999), 21–106.

55. See, for example, Cavendish, "The Second Part of Bell Campo," in Shaver, *Convent of Pleasures and Other Plays*, 167, and "The She-Anchoret," in *NP*, 287–357.

56. Daston and Park, *Wonders and the Order of Nature*, 176–77.

57. Steven Shapin has argued that the Royal Society was desperately, and largely unsuccessfully, trying to articulate a new ideal of the gentleman-scholar in the early 1660s (Steven Shapin, "A Scholar and a Gentleman: The Problematic Identity of the Scientific Practitioner in Early Modern England," *History of Science* 29 (1991): 279–327, and Shapin, *A Social History of Truth: Civility and Science in Seventeenth-Century England* (Chicago: University of Chicago Press, 1994).

58. Findlen writes in *Possessing Nature*, "Experiencing nature became part of the aesthetic production of the late Renaissance and Baroque court, where knowledge was simply another form of display; concomitantly it also drew upon an urban culture that placed spectacle, in all its forms, at its public center" (201).

59. For example, see *OEP*, 11.

60. *SL*, 206.

61. Shapin and Schaffer, *Leviathan and the Air-Pump*, 30–34.

62. Robert Boyle, *Some Considerations Touching the Usefulness of Natural Philosophie*, in Boyle, *The Works*, ed. Thomas Birch, with an intro. by Douglas McKie (Hildesheim, Germany: Georg Olms Verlagsbuchhandlung, 1966), 30. Findlen, "Between Carnival and Lent," argues that the Royal Society was affected by the religious

and cultural effort to destroy popular culture, symbolized in the triumph of Lent over Carnival (263).

63. Robert Boyle, preface to *Experiments and Considerations Touching Colours* (facs. 1664 ed., New York: Johnson Reprint Corp., 1964), n.p. In *Observations* (59–73), Cavendish refers to several experiments and claims Boyle had made in *Experiments*.

64. Thomas Sprat, *History of the Royal Society*, ed. Jackson I. Cope and Harold Whitmore Jones (St. Louis: Washington University Studies; London: Routledge & Kegan Paul, 1959), 90–91.

65. Ibid., 214–15.

66. Pepys, *Diary*, 8:163.

67. John Evelyn, *The Diary of John Evelyn*, ed. E. S. de Beer (London: Oxford University Press, 1959), 11:31. Mary Evelyn to Ralph Bohun, April 1667, quoted in Frances Harris, "Living in the Neighbourhood of Science: Mary Evelyn, Margaret Cavendish and the Greshamites," in Hunter and Hutton, *Women, Science and Medicine*, 198–99. Harris, in her discussion of Mary Evelyn's reactions to Cavendish and her position in relation to learning and science, also emphasizes the performative nature of Cavendish's visit to the Royal Society: "Most of those who witnessed it were prepared to enjoy the show" (198). However, Harris supports the traditional interpretation of Cavendish's visit.

68. Pepys, *Diary*, 8:243.

69. Harris, "Living in the Neighbourhood of Science," 208.

70. Quoted in Michael Hunter, *Science and Society in Restoration England* (Cambridge: Cambridge University Press, 1981), 34–35. Hunter discusses the ways the Royal Society attempted to distinguish itself from other "social clubs."

71. Dudley North to his father, 1667, Bodleian Library MS North c. 4 146r. I wish to thank Katie Whitaker for this reference.

72. Pepys, *Diary*, 8:243.

73. Evelyn, *Diary*, 33.

74. The *Oxford English Dictionary* defines *antic* as "absurd from fantastic incongruity; grotesque, bizarre, uncouthly, ludicrous." As a noun, it is "a grotesque pageant or theatrical performance."

75. Thomas Birch, *The History of the Royal Society for Improvement of Natural Knowledge from its First Rise* (London, 1756–57; facs., New York: Johnson Reprint Corp., 1968), 2:175–78.

76. Evelyn, *Diary*, 33; Mintz, "Duchess of Newcastle's Visit to the Royal Society," 176.

77. Pepys, *Diary*, 8:243.

78. Birch, *History of the Royal Society*, 177. The list of experiments Birch gives is not quite the same as those described by Pepys.

79. Pepys, *Diary*, 8:243.

80. Whitaker, *Mad Madge*, 300.

81. Cavendish, "The Hunting of the Hare," in *P & F*, 120.

82. These definitions come from the *Oxford English Dictionary*.

83. Boyle, preface to *Experiments*, n.p.

84. Cavendish, "Further Observations," in *BW*, 20–22.

85. Birch, *History of the Royal Society*, 250, and Michael Hunter, *Establishing the New Science: The Experience of the Early Royal Society* (Woodbridge, Suffolk, England: Boydell Press, 1989), 167, 171.

86. *GNP*, n.p.

87. Quoted in Hunter, *Science and Society*, 131.

## 2: CAVENDISH'S EARLY ATOMISM

1. Cavendish, "To Natural Philosophers," in *P & F*, n.p.

2. Cavendish, "The Poetresses Petition," in *P & F*, n.p.

3. Cavendish, "An excuse for so much writ upon my Verses," in *P & F*, n.p.

4. Sara Mendelson and Patricia Crawford, *Women in Early Modern England* (Oxford: Oxford University Press, 2000), 28. Cavendish was childless herself and thus failed in what was considered the major obligation of married women. Katie Whitaker argues that her infertility was due to Newcastle's impotence, but in social terms barrenness was almost always blamed on the wife (*Mad Madge: The Extraordinary Life of Margaret Cavendish, Duchess of Newcastle, the First Woman to Live by Her Pen* [New York: Basic Books, 2002], 100–101).

5. Robert Kargon, *Atomism in England from Hariot to Newton* (Oxford: Clarendon Press, 1966), 73. For other interpretations of Cavendish's atomism, see Lisa T. Sarasohn, "A Science Turned Upside Down: Feminism and the Natural Philosophy of Margaret Cavendish," *Huntington Library Quarterly* 47 (1984): 289–307; John Rogers, *The Matter of Revolution, Science, Poetry, and Politics in the Age of Milton* (Ithaca, NY: Cornell University Press, 1996), 184; and Stephen Clucas, "The Atomism of the Cavendish Circle: A Reappraisal," *Seventeenth Century* 9 (1994): 247–73.

6. On Gassendi, see Lisa T. Sarasohn, *Gassendi's Ethics: Freedom in a Mechanistic Universe* (Ithaca, NY: Cornell University Press, 1986), and Margaret J. Osler, *Divine Will and the Mechanical Philosophy: Gassendi and Descartes on Contingency and Necessity in the Created World* (Cambridge: Cambridge University Press, 1994).

7. On the relationship between skepticism and fideism, especially how it pertains to Montaigne, see Richard H. Popkin, *The History of Scepticism from Erasmus to Descartes*, rev. ed. (New York: Harper & Row, 1964), 45–66.

8. Cavendish, "To the Reader," in *P & F*, n.p.

9. Cavendish, *Orations of Divers Sorts, Accommodated to Divers Places* (1662), 322–23.

10. *PF*, 93–94; *PPO* 1655, 172; *NP*, 290. A slightly different usage can be found in "The Presence," in *Plays Never Before Printed*, 34.

11. www.newadvent.org/cathen/06068b.htm, accessed January 19, 2009.

12. *TR*, 51.

13. Cavendish, "The Epistle Dedicatory: To Sir Charles Cavendish, My Noble Brother-in-Law," in *P & F*, n.p.

14. In Cavendish's memoir, *A True Relation of my Birth, Breeding, and Life*, she described Sir Charles as "nobly generous, wisely valiant, naturally civill, honestly

kind, truly loving, vertuously temperate; his promise was like a fixt degree, his words were destiny" (*TR*, 50).

15. *P & F*, 240.

16. Cavendish, "To Naturall Philosophers," in *P & F*, n.p.

17. Cavendish, "To Learned Philosophers," in *P & F*, n.p.

18. Cavendish, "A Dedication to Fame," in *PF*, n.p.

19. Cavendish, "To Naturall Philosophers," in *P & F*, n.p.

20. Cavendish, "To the Reader," in *PPO* 1655, n.p.

21. Cavendish, "An Elegy," in *PL*, 85.

22. Virginia Woolf, *The Second Common Reader* (New York: Harcourt, Brace, 1932), 76.

23. *WO*, 22. Many of the ideas expressed in *Poems, and Fancies, Philosophicall Fancies,* and *Physical and Philosophical Opinions* appear in this text, which is a collection of short essays.

24. *P & F*, 1.

25. *P & F*, 1–2.

26. Cavendish, "My Lord," in *WO*, n.p.

27. Ibid.

28. *P & F*, 2–3.

29. *Oxford English Dictionary.*

30. *P & F*, 2.

31. Ibid.

32. *P & F*, 3–4.

33. *P & F*, 4.

34. *P & F*, 6.

35. In antiquity, Democritus had argued that all is determined by the mechanistic action of the atoms, while Epicurus has inserted a principle of indeterminism into the action of atoms: an unexpected and uncaused swerve that freed matter from necessity. See Sarasohn, *Gassendi's Ethics*, 136–41.

36. *P & F*, 14.

37. *P & F*, 19.

38. *P & F*, 9.

39. *P & F*, 12.

40. *P & F*, 31.

41. *P & F*, 40.

42. *P & F*, 4.

43. *P & F*, 41.

44. Ibid.

45. *P & F*, 41.

46. *P & F*, 42.

47. On Epicurean religion, see J. M. Rist, *Epicurus: An Introduction* (Cambridge: Cambridge University Press, 1972).

48. Sir Charles Cavendish to John Pell, September 10, 1645, and September 6, 1650, British Museum Additional Mss. 4280.

49. *Letters and Poems in Honour of the Incomparable Princess, Margaret, Dutchess of Newcastle* (London, 1676), 145–48.

50. On Charleton's life and writings, see Kargon, *Atomism in England*, 77–92.

51. Walter Charleton, *The Darkness of Atheism Dispelled by the Light of Nature: A Physico-theologicall Treatise* (London, 1652), 44.

52. On Charleton's voluntarism, see Margaret J. Osler, "Descartes and Charleton on Nature and God," *Journal of the History of Ideas* 40 (1979): 45–56.

53. Nina Rattner Gelbart, "The Intellectual Development of Walter Charleton," *Ambix* 18 (1971): 149–68. Gelbart has argued that Charleton retained many vitalist elements of the chemical philosophy in his later atomism: "Thus, try as he might, Charleton could not become a Mechanist of the deepest dye (168)."

54. Charleton, *Darkness of Atheism*, 43–44.

55. *Letters and Poems*, 143–44.

56. Cavendish, "An Epiloge to my Philosophical Opinions," in *PPO* 1655, n.p.

57. Douglas Bush, *Mythology and the Renaissance Tradition in English Poetry* (New York: Norton, 1932; rev. ed., 1963); Keith Thomas, *Religion and the Decline of Magic* (New York: Charles Scribner's Sons, 1971); Martin Ingram, "From Reformation to Toleration: Popular Religious Cultures in England, 1540–1690," in *Popular Culture in England, c. 1500–1850*, ed. Tim Harris (New York: St. Martin's Press, 1995), 95–123.

58. *Letters and Poems*, 144.

59. *P & F*, 43–44.

60. *P & F*, 44–45.

61. *P & F*, 187.

62. *P & F*, 174.

63. *P & F*, 175.

64. *P & F*, 187, 16.

65. *P & F*, 25, 188.

66. William Shakespeare, *Romeo and Juliet*, in *The Arden Shakespeare Complete Works*, ed. Richard Proudfoot, Ann Thompson, and David Harold Jenkins (Walton-on-Thames, Surrey, England: Thomas Nelson and Sons, 1989), 1012.

67. Katharine Briggs, *The Anatomy of Puck: An Examination of Fairy Beliefs among Shakespeare's Contemporaries and Successors* (London: Routledge & Kegan Paul, 1959), 56–70. Briggs is a folklorist whose book is a repository of information about fairies but must be used with caution by the historian. A more sophisticated treatment of the relationship between atoms and fairies in the early Stuart period can be found in Reid Barbour, *English Epicures and Stoics: Ancient Legacies in Early Stuart Culture* (Amherst: University of Massachusetts Press, 1998), who discusses the political implications of a mythologized and miniaturized fairy court for the Stuart monarchy (36–49).

68. Cavendish's praise of Shakespeare appears in *SL*, 130–31.

69. Quoted in Briggs, *Anatomy of Puck*, 32. The fact that fairies were sometimes associated with witches and their familiars is evident in Scot's treatise. See also Regina Buccola, *Fairies, Fractious Old Women, and the Old Faith: Fairy Lore in Early*

*Modern British Drama and Culture* (Cranbury, NJ: Associated University Presses, 2006).

70. Barbour, *English Epicures and Stoics*, 38.

71. Lucretius, *On the Nature of the Universe (De Rerum Natura)*, trans. James H. Mantinband (New York: Frederick Ungar, 1968), 37.

72. Emma E. L. Rees, "'Sweet Honey of the Muses': Lucretian Resonance in Poems and Fancies," *In-between: Essays and Studies in Literary Criticism* 9 (2000): 3–16.

73. John Donne, *Elegy XIV*, in *The Complete English Poems* (London: Penguin, 1986), 114, and Barbour, *English Epicures and Stoics*, 38.

74. Briggs, *Anatomy of Puck*, argues that the popular writers appealed to "the Court as well as the Country," both the traditionally educated and the newly literate, in the late sixteenth and seventeenth centuries (6).

75. Cavendish, "The Sociable Companions," in *Plays Never Before Printed*, 46.

76. *P & F*, 178.

77. *P & F*, 175.

78. *P & F*, 17.

79. *P& F*, 63–64. This passage is also directed against those who believe in witches and immaterial spirits, an attack that continues in *Philosophical Letters*.

### 3: THE LIFE OF MATTER

1. *Philosophicall Fancies* was printed separately in 1653 and reprinted as part 1 of *Philosophical and Physical Opinions* in 1655.

2. William R. Newman, *Atoms and Alchemy: Chymistry and the Experimental Origins of the Scientific Revolution* (Chicago: University of Chicago Press, 2006), traces the role atomism, Aristotelian matter theory, and alchemy played in the scientific revolution.

3. Cavendish, "An Epiloge to My Philosophical Opinions," in *PPO* 1655, n.p.

4. See John Henry, "Occult Qualities and the Experimental Philosophy: Active Principles in Pre-Newtonian Matter Theory," *History of Science* 24 (1986): 335–81; Margaret J. Osler, "How Mechanical Was the Mechanical Philosophy?" in *Late Medieval and Early Modern Corpuscular Matter Theories*, ed. Cristoph Lüthy, John E. Murdoch, and William R. Newman (Leiden: Springer, 2001), 423–39; and Antonio Clericuzio, *Elements, Principles and Corpuscles: A Study of Atomism and Chemistry in the Seventeenth Century* (Dordrecht: Kluwer, 2000).

5. *PF*, 15.

6. Cavendish, "A Condemning Treatise of Atomes," in *PPO* 1655, n.p.

7. Peter Anstey writes that Boyle felt, "The Aristotelians are constantly appealing to a *horror vacui*, to innate appetites, sympathies and antipathies. But worse, a philosophic analysis of these terms reveal far from being metaphorical for the Aristotelians, they are actually real qualities attributable to material objects. The sort of late scholastic accounts of the behavior of matter Boyle criticized is ultimately a form of panpsychism" (Anstey, "Boyle against Thinking Matter," in Lüthy, Murdoch, and

Newman, *Late Medieval and Early Modern Corpuscular Matter Theories*, 501). See also Antonio Clericuzio, "Gassendi, Charleton and Boyle on Matter and Motion," in Lüthy, Murdoch, and Newman, *Late Medieval and Early Modern Corpuscular Matter Theories*, 467–82.

8. Cavendish, "To the Reader," in *PF*, n.p.

9. Most scholars who have addressed the question of when Cavendish adopted vitalistic materialism usually trace it to the 1655 *Philosophical and Physical Opinions*. See, for example, Stephen Clucas, "The Duchess and the Viscountess: Negotiations between Mechanism and Vitalism in the Natural Philosophies of Margaret Cavendish and Anne Conway," *In-between: Essays and Studies in Literary Criticism* 9 (2000): 125–28.

10. *PF*, 12–13.

11. *PF*, 49.

12. *PF*, 12.

13. Thomas Hobbes, *Leviathan*, ed. Richard Tuck (Cambridge: Cambridge University Press, 1991), 87.

14. *PF*, 13.

15. *PF*, 15.

16. *PPO* 1655, 96.

17. *PF*, 18–19.

18. *PF*, 17.

19. *PF*, 30.

20. *PF*, 31.

21. *PF*, 39.

22. *PF*, 34.

23. *PF*, 39.

24. *PF*, 42.

25. *PF*, 8–9.

26. There is a complex history of particulate alchemy, various elements of which are reflected in Cavendish's matter theory, although probably through osmosis rather than direct borrowing. Nina Rattner Gelbart argues that in his early Helmontian texts Charleton embraced a form of corpuscular vitalism: the *minima naturalia*, which postulated that the world was made up of qualitatively different atoms. This conception can be traced back to Aristotle and continued through the early seventeenth century in the thinking of the physician and alchemist Daniel Sennert (Gelbart, "The Intellectual Development of Walter Charleton," *Ambix: The Journal for the Study of Alchemy and Early Chemistry* 18 [1971]: 129–57). William R. Newman finds even stronger evidence that aspects of Sennert's corpuscularism reflected an Aristotelian tradition that saw minute particles as the building blocks of nature and was therefore similar to Democritean atomism (Newman, "Corpuscular Alchemy and the Tradition of Aristotle's *Meteorology,* with Special Reference to Daniel Sennert," *International Studies in the Philosophy of Science* 15 [2001]: 145–53). Stephen Clucas, "The Atomism of the Cavendish Circle: A Reappraisal," *Seventeenth Century* 9 (1994): 251–52, identifies a tradition he calls "neo-atomism" that links the thought of

Sennert, Joachim Jungius (an acquaintance of Newcastle's in Paris), Van Helmont, Charleton, and Cavendish. Van Helmont is one of the philosophers Cavendish attacked in *Philosophical Letters* (1664).

27. *PF*, 12–13, 18. Vitriol is defined by Betty Jo Dobbs as "any sulfate, but probably that of iron or copper" (Dobbs, *The Foundations of Newton's Alchemy or "The Hunting of the Greene Lyon"* [Cambridge: Cambridge University Press, 1975], xiii).

28. *PF*, 39.

29. Paracelsus, *Paracelsus: Selected Writings*, ed. Jolande Jacobi (Princeton: Princeton University Press, 1951), 257.

30. *PF*, 30.

31. *PF*, 53.

32. Allen B. Debus, *The Chemical Philosophy: Paracelsian Science and Medicine in the Sixteenth and Seventeenth Centuries* (New York: Science History Publications, 1977), 2:340. See also Walter Pagel, *Joan Baptista Van Helmont: Reformer of Science and Medicine* (Cambridge: Cambridge University Press, 1982), 96–97.

33. Pagel, *Joan Baptista Van Helmont*, 40.

34. Quoted in ibid., 97.

35. Walter Charleton, "Prolegemena," in Johannes Baptista Van Helmont, *Ternary of Paradoxes*, translated, illustrated, and amplified by Walter Charleton (London, 1650).

36. Stephen Clucas, "The Atomism of the Cavendish Circle," 226, suggests that atomic ideas of Cavendish and Hobbes are inclusive of many different theories of particulate matter and were not primarily inspired by Gassendi's Epicurean atomism.

37. Charleton, "Prolegomena." Charleton probably found the metaphor of the bees in Van Helmont. See Pagel, *Joan Baptista Van Helmont*, 21–22.

38. *PF*, 54–55

39. *PF*, 54–56.

40. *PF*, 62.

41. Walter Charleton, *The Darkness of Atheism dispelled by the light of Nature a physico-theological treatise* (London, 1652).

42. *PF*, 93.

43. Ibid.

44. *PF*, 94.

45. *PF*, 76.

46. Cavendish, "To the Two Universities," in *PPO* 1655, n.p.

47. Cavendish, "An Epilogue to my Philosophical Opinions," in *PPO* 1655, n p.

48. Cavendish, "A Condemning Treatise of Atomes," in *PPO* 1655, n.p.

49. Ibid.

50. Cavendish, "An Epistle to the Reader, for my Book of Philosophy," in *PPO* 1655, n.p.

51. *P & F*, 18.

52. Cavendish, "The Text to my Natural Sermon," in *PPO* 1655, n.p.

53. Cavendish, "Epistle to My Reader," in *PPO* 1655, n.p.

54. Charleton, *Physiologia Epicuro-Gassendi-Charletoniana, or a Fabrick of Science Natural Upon the Hypothesis of Atoms, Founded by Epicurus, Repaired by Petrus Gassendus. Augmeted by Walter Charleton* (London, 1654; repr., New York: Johnson Reprint, 1966), 111–12.

55. Ibid., 113.

56. Ibid., 126.

57. Voluntarism is the theological argument that asserts that God's power is an aspect of his will and that he can do anything he wants, including changing the laws of nature. On Charleton's voluntarism, see Margaret J. Osler, "Descartes and Charleton on Nature and God," *Journal of the History of Ideas* 40 (1979): 445–56.

58. *PPO* 1655, 30.

59. Cavendish, "An Epistle to my Readers," in *PPO* 1655, n.p.

60. Cavendish, "The Text to my Natural Sermon," in *PPO* 1655, n.p. I have retained the typeset of the text, to emphasize the litany of Cavendish's natural philosophy.

61. *SL*, 98.

62. Cavendish, "Another Epistle to the Reader," in *PPO* 1663, n.p.

63. *PPO* 1655, 47.

64. Ibid., 37.

65. All of these definitions are taken from the 2005 *Oxford Dictionary Online*.

66. *PPO* 1655, 34.

67. Ibid., 47–48.

68. Ibid., 50.

69. Ibid., 103–4.

70. Ibid., 104–5.

71. Ibid., 32.

72. Ibid., 41.

73. Ibid.

74. Ibid., 104.

75. Ibid., 107.

76. Ibid., 110–16. See Kouren Michaelian, "Margaret Cavendish's Epistemology," *British Journal for the History of Science* 17 (2009): 31–53, for a sophisticated analysis of Cavendish's theory of patterning.

77. David Knowles, *The Evolution of Medieval Thought*, 2nd ed. (London: Longman, 1988), 11–13, 187–98.

78. *PPO* 1655, 42.

79. Ibid., 42–43.

80. Ibid., 97–98.

81. Ibid., 52.

82. *P & F*, 46–47.

83. *PPO* 1655, 51–52.

84. Ibid., 48, 51.

85. Cavendish, "Epistle to my Readers," in *PPO* 1655, 172.

## 4: THE IMAGINATIVE UNIVERSE OF *NATURES PICTURES*

1. For a discussion of Cavendish's debt to earlier fantasy writers, especially Cyrano de Bergerac, see Line Cottegnies, "Margaret Cavendish and Cyrano de Bergerac: A Libertine Subtext for Cavendish's Blazing World (1660)?" *Bulletin de la Societé d'Études Anglo-Americaines des xviie et xviiie Siècles* 54 (2002): 165–85.

2. Cavendish, "To the Reader," in *NP,* n.p.

3. William Newcastle, "A Copy of Verses to the Lady Machioness of Newcastle," in *NP,* n.p.

4. Cavendish, "An Epistle to My Readers," in *NP,* n.p.

5. Cavendish, "To My Readers," in *NP,* n.p.

6. *NP,* 393.

7. Kate Lilley makes this point in *The Blazing World and Other Writings,* ed. Lilley (London: Pickering, 1992; repr., London: Penguin, 1994), xxiii. Also see Sujata Iyengar's discussion of race and color in *Assaulted and Pursued Chastity* in "Royalist, Romancist, Racialist: Rank, Gender, and Race in the Science and Fiction of Margaret Cavendish," *ELH* 69 (2002). 649–72, 657–58.

8. Cavendish, "The She-Anchoret," in *NP,* 288, 356.

9. Cavendish, "To the Two Universities," in *PPO* 1655, n.p.

10. Margaret L. King, "Book-Lined Cells: Women and Humanism in the Early Italian Renaissance," in *Beyond Their Sex: Learned Women in the European Past,* ed. Patricia H. Labalme (New York: New York University Press, 1980), 74.

11. *TR,* 57.

12. The different types include the following: natural philosophers, physicians, moral philosophers, Holy Fathers of the Church, judges, barristers and orators, statesmen, tradesmen or citizens, housekeepers and masters of families, etc., married men and their wives, nurses and their nurse-children, widowers and widows, virgins, lovers, poets, aged persons, soldiers, historians.

13. Cavendish, "The She-Anchoret," 289.

14. Ibid., 307.

15. Ibid., 290.

16. Cavendish, "The Expression of the Doubts and Curiosity of Man's Minde," in *NP,* 74.

17. Cavendish, "The She-Anchoret," 297.

18. Ibid., 318–19.

19. Carolyn Merchant, *The Death of Nature: Women, Ecology, and the Scientific Revolution* (San Francisco: Harper & Row, 1980), 127.

20. Cavendish, "The Vulgar Fights," in *NP,* n.p.

21. Natalie Zemon Davis, *Society and Culture in Early Modern France: Eight Essays* (Stanford, CA: Stanford University Press, 1975), 130–31.

22. Cavendish, "The She-Anchoret," 319.

23. *P & F,* 16.

24. Cavendish, "The She-Anchoret," 319.

25. *GNP,* 265–89.

26. Cavendish, "The She-Anchoret," 319.

27. Ibid.

28. *NP*, 156.

29. Ibid.

30. Ibid.

31. *NP*, 74–75.

32. Patricia Crawford, *Women and Religion in Early Modern Europe* (London: Routledge, 1993), 78.

33. Marina Leslie, "Evading Rape and Embracing Empire in Margaret Cavendish's *Assaulted and Pursued Chastity*," in *Menacing Virgins: Representing Virginity in the Middle Ages and Renaissance*, ed. Kathleen Coyne Kelly and Marina Leslie (Newark: University of Delaware Press; London: Associated University Presses, 1999), 179.

34. *NP*, 311.

35. *NP*, 177.

36. Quoted in Christopher Hill, "Irreligion in the 'Puritan' Revolution," in *Radical Religion in the English Revolution*, ed. J. F. McGregor and B. Reay (Oxford: Oxford University Press, 1984), 206. The French were equally concerned about atheism. See Alan Charles Kors, *Atheism in France, 1650–1729*, vol. 1: *The Orthodox Sources of Disbelief* (Princeton: Princeton University Press, 1990), 17–43.

37. Michael Hunter, *Science and the Shape of Orthodoxy: Intellectual Change in Late-Seventeenth Century Britain* (Woodbridge, Suffolk, England: Boydell Press, 1995), 229; Barbara J. Shapiro, *Probability and Certainty in Seventeenth-Century England: A Study of the Relationships between Natural Science, Religion, History, Law, and Literature* (Princeton: Princeton University Press, 1983), 82–83.

38. On Epicurean ethics, see Lisa T. Sarasohn, *Gassendi's Ethics: Freedom in a Mechanistic Universe* (Ithaca, NY: Cornell University Press, 1986), 51–57.

39. Cavendish, "The She-Anchoret," 317.

40. Hobbes, *Leviathan*, ed. Richard Tuck (Cambridge: Cambridge University Press, 1991), 270.

41. Quoted in Samuel I. Mintz, *The Hunting of Leviathan: Seventeenth-Century Reactions to the Materialism and Moral Philosophy of Thomas Hobbes* (Cambridge: Cambridge University Press, 1962), 109.

42. Cavendish, "The She-Anchoret," 289. This passage is reproduced with some differences in "Assaulted and Pursued Chastity," another romance in *Natures Pictures*.

43. Cavendish, "Assaulted and Pursued Chastity," in *NP*, 239.

44. Cavendish, "The Expression of the Doubts and Curiosities of Man's Minde," 74.

45. This conclusion is echoed by the She-Anchoret: "Then they asked her what difference there is between the soul and the minde. She answered, as much difference as there is betwixt flame, and grosser part of fire; for, said she, the soul is onely the pure part of the minde" (Cavendish, "The She-Anchoret," 317).

46. Cavendish, "The Expression of the Doubts and Curiosities of Man's Minde," 74.

47. Cavendish, "The She-Anchoret," 318.

48. On religious enthusiasm during the English Civil War, see J. F. McGregor, "The Baptists: Font of all Heresies" and "Seekers and Ranters," in McGregor and

Reay, *Radical Religion*, 57–62, 121–39. On the Royal Society and enthusiasm, see Steven Shapin and Simon Schaffer, *Leviathan and the Air-Pump: Hobbes, Boyle and the Experimental Life* (Princeton: Princeton University Press, 1985), 72–76.

49. Cavendish, "The She-Anchoret," 320.

50. For a full account of this dispute, see Arnold A. Rogow, *Thomas Hobbes: Radical in the Service of Reaction* (New York: Norton, 1986), 178–83.

51. Cavendish, "The She-Anchoret," 290.

52. Thomas Hobbes, "Of Liberty and Necessity," in *The English Works of Thomas Hobbes of Malmesbury*, ed. Sir William Molesworth (London: John Bohn, 1840; repr., Germany: Scientia Verlag Aalen, 1966), 4:246.

53. Ibid.

54. Cavendish, "The She-Anchoret," 318.

55. Hobbes first develops this concept in *Elements of Law* (1640), also dedicated to Newcastle. It is developed at length in "Of Liberty and Necessity," where he sums up the argument in the following manner: "Liberty is the absence of all the impediments to action that are not contained in the nature and intrinsical quality of the agent" (273). On Hobbes's determinism, see Sarasohn, *Gassendi's Ethics*, 126–28.

56. Hobbes, "Of Liberty and Necessity," 260.

57. Cavendish, "The She-Anchoret," 318.

58. Ibid.

59. On Hobbes's psychology, see Bernard Gert, "Hobbes's Psychology," in *The Cambridge Companion to Hobbes*, ed. Tom Sorell (Cambridge: Cambridge University Press, 1996), 157–74, and Sarasohn, *Gassendi's Ethics*, 118–36. There is another treatise, which may have been written by Hobbes, called "A Short Tract on First Principles," which also contains his psychological ideas. Some critics credit this work to Hobbes, and others to Robert Payne, who was Newcastle's chaplain in the 1630s, or even perhaps to Sir Charles Cavendish himself. Whoever wrote it, it clearly reflects the psychological theories advocated by members of the Cavendish Circle in the early 1630s. See Thomas Hobbes, "A Short Tract on First Principles," in *The Elements of Law, Natural and Political*, ed. Ferdinand Tönnies (London: Cambridge University Press, 1920), 208–9; Frithiof Brandt, *Thomas Hobbes' Mechanical Conception of Nature* (Copenhagen: Levin & Munksgaard, 1928), 48–50; and Richard Tuck, "Hobbes and Descartes," in *Perspectives on Thomas Hobbes*, ed. G. A. J. Rogers and Alan Ryan (Oxford: Clarendon Press, 1988).

60. René Descartes, "The Passions of the Soul," in *The Philosophical Writings of Descartes*, vol. 1, edited by John Cottingham, Robert Stoothhoff, and Dugald Murdoch (Cambridge: Cambridge University Press, 1985), 335.

61. Kourken Michaelian, "Margaret Cavendish's Epistemology," *British Journal for the History of Philosophy* 17 (2009): 31–53. Michaelian's analysis of Cavendish's epistemology is very close to mine. He argues that "her answer to a (less familiar) question about what knows—her surprising answer, in brief, is: everything " (32).

62. Cavendish, "The She-Anchoret," 308.

63. Ibid., 310–11.

64. Ibid., 308.

65. Ibid., 309.

66. Ibid.

67. Hobbes, *Leviathan*, 18.

68. Cavendish, "The Traveling Spirit," in *NP*, 146.

69. On witchcraft and More and Glanvil, see A. Rupert Hall, *Henry More and the Scientific Revolution* (Cambridge: Cambridge University Press, 1997), 138–43. For More and Glanvil, the existence of witches was necessary because of their doctrine of immaterial substances, which is the opposite of Cavendish's point of view. But even Hobbes acknowledged that there were people who imagined they were witches; although they had no real supernatural powers, they could do mischief and create disorder in the state (*Leviathan*, 18). Newcastle and Hobbes discussed the possible existence of witches while the Cavendishes were in Paris. Newcastle argued that witches were simply deluded into believing they had a contract with the Devil, and they were dreaming when they thought they flew or changed shapes. Nevertheless, Cavendish concluded, "My Lord doth not count this opinion of his so universal, as if there were none but imaginary witches" (*Life* 1916, 135–36.)

70. Keith Thomas, *Religion and the Decline of Magic* (New York: Scribner, 1971), 608–9; K. M. Briggs, *The Anatomy of Puck: An Examination of Fairy Beliefs among Shakespeare's Contemporaries and Successors* (London: Routledge & Kegan Paul, 1959), 46–47.

71. Cavendish, "The Traveling Spirit," 146.

72. Ibid., 147.

73. Ibid., 146. Whether Cavendish is a mortalist, like Hobbes and Milton, who believed that the body died until resurrected with the soul at the second coming, is unclear here. On mortalism, see Hill, "Irreligion, " 199–204.

74. Cavendish, "The Speculators," in *NP*, 149.

75. The discussion of optics in "The She-Anchoret" focuses on the question of whether cats can see in the dark (298–99).

76. Cavendish, "The Speculators," 149–53.

77. Cavendish, "Assaulted and Pursued Chastity," 231.

78. Mary Baine Campbell, *Wonder and Science: Imagining Other Worlds in Early Modern Europe* (Ithaca, NY: Cornell University Press, 1999), 205.

79. On the description of Amerindians and its relationship to traditional European images of other lands, see Anthony Pagden, *European Encounters with the New World, from Renaissance to Romanticism* (New Haven: Yale University Press, 1992), and John F. Moffitt and Santiago Sebastián, *O Brave New People: The European Invention of the American Indian* (Albuquerque: University of New Mexico Press, 1996).

80. Quoted in Bernard W. Sheehan, *Savagism and Civility: Indians and Englishmen in Colonial Virginia* (Cambridge: Cambridge University Press, 1980), 188.

81. Cavendish, "Assaulted and Pursued Chastity," in Bowerbank and Mendelson, *Paper Bodies*, 63.

82. On hybrid beasts, see Arnold I. Davidson, "The Horror of Monsters," in *The Boundaries of Humanity: Humans, Animals, Machines*, ed. James J. Sheehan and Morton Sosna (Berkeley: University of California Press, 1991), 36–67, and Peter Costello,

*The Magic Zoo: The Natural History of Fabulous Beasts* (New York: St. Martin's Press, 1979).

83. On the role of women in English radical religion, see Crawford, *Women and Religion,* 119–59.

84. Cavendish, "Assaulted and Pursued Chastity," 239.

85. Ibid., 239.

86. Cavendish, "The Expression of the Doubts and Curiosities of Man's Minde," 70.

87. Cavendish, "A Condemning Treatise of Atomes," in *PPO* 1655, n.p.

88. On Cavendish and Stoicism, see Eileen O'Neill in her introduction to the first modern edition of Cavendish's *Observations upon Experimental Philosophy* (Cambridge: Cambridge University Press, 2001), x—xxxvi.

89. Cavendish, "The She-Anchoret," 317–18.

90. Ibid., 317.

## 5: THE POLITICS OF MATTER

1. *Letters and Poems,* 158.

2. Cavendish, *Life* 1916, 5.

3. Ibid., 6.

4. Cavendish, "Epistle to the Reader," in *PPO* 1663, n.p.

5. On the changes between the two editions of *Philosophical and Physical Opinions,* see Whitaker, *Mad Madge,* 251–54. The text of the work was more than doubled, going from 172 to 458 pages.

6. On Cavendish's politics, see David Norbrook, "Margaret Cavendsih and Lucy Hutchinson: Identity, Ideology, and Politics," *In-between: Essays and Studies in Literary Criticism* 9, nos. 1–2 (2000): 179–203; Emma E. L. Rees, *Margaret Cavendish: Gender, Genre, Exile* (Manchester: Manchester University Press, 2003), 44–46, 1–17; John Rogers, *The Matter of Revolution: Science, Poetry, and Politics in the Age of Milton* (Ithaca, NY : Cornell University Press, 1996), 1–16, 195–97; Hilda Smith, " 'A General War amongst the Men . . . but None amongst the Women': Political Differences between Margaret and William Newcastle," in *Politics and the Political Imagination in Later Stuart Britain: Essays Presented to Lois Green Schwoerer,* ed. Howard Nenner (Rochester, NY: Rochester University Press, 1997), 143–60; and Mihoko Suzuki, *Subordinate Subjects: Gender, the Political Nation, and Literary Form in England, 1588–1688* (Aldershot, England: Ashgate, 2003).

7. On atomism, politics, and Hobbes, see Lisa T. Sarasohn, "Motion and Morality: Pierre Gassendi, Thomas Hobbes and the Mechanical World-View," *Journal of the History of Ideas* 46 (1985): 363–79; Thomas A. Spragens, *The Politics of Motion: The World of Thomas Hobbes* (Lexington: University of Kentucky Press, 1973). Spragens writes, "Man is himself in Hobbes's view one part of the whole of nature, and he could therefore justifiably be considered as sharing the fundamental properties of the natural order" (129). Spragens's title inspires the title of this chapter. See also Michael Verdon, "On the Laws of Physical and Human Nature: Hobbes' Physical and Social Cosmologies," *Journal of the History of Ideas* 43 (1982): 656–59.

8. Cavendish, "A Condemning Treatise on Atomes," in *PPO* 1655, n.p.

9. Cavendish, "To the Reader," in *PPO* 1663, n.p.

10. Ibid..

11. Rogers makes just this argument in *The Matter of Revolution* (1–16, 195–97).

12. Cavendish, preface to *PPO* 1663, n.p. Of course, unlike Parmenides, Cavendish's matter possesses motion.

13. See, for example, *PPO* 1663, 42, 45–46, 87.

14. Ibid., 2–3.

15. Ibid., 14.

16. *PL*, 493–94.

17. *PPO* 1663, 3.

18. Ibid., 10–11.

19. Ibid., 75.

20. Ibid., 278.

21. Cavendish, *Orations of Divers Sorts, Accommodated to Divers Places* (1662), in *Margaret Cavendish: Political Writings*, ed. Susan James (Cambridge: Cambridge University Press, 2003), 134–36 194–95.

22. *P & F*, 199.

23. *P & F*, 200.

24. *WO*, 205–13.

25. *PPO* 1655, 32.

26. Ibid., 44.

27. Cavendish, preface to *Orations of Divers Sorts, Accommodated to Divers Places* (London, 1662), n.p.

28. Cavendish, *Orations*, 110.

29. Ibid., 277.

30. Ibid., 286–87.

31. Ibid., 70.

32. Ibid., 124, 262.

33. Ibid., 280.

34. Ibid.

35. Ibid., 120, 123, 130, 283.

36. Ibid., 116–17.

37. Cavendish, *Orations*, 87. Compare this statement with the beginning of the Digger proclamation, *The True Leveller Standard Advanced* (1649), "In the beginning of Time, the great Creator Reason, made the Earth to be a Common Treasury, to preserve Beasts, Birds, Fishes, and Man, the lord that was to govern this Creation" (Gerard Winstanley, *The True Leveller Standard Advanced* [London, 1649], n.p.).

38. Cavendish, *Orations*, 86–87.

39. Ibid., 88.

40. "A Cause Pleaded at the Bar before Judges, concerning Theft," in *Orations*, 88.

41. Ibid., 69.

42. Ibid., 222.

43. Ibid., 224.

44. Cavendish, *Orations of Divers Sorts, Accommodated to Divers Places* (1662), in *Paper Bodies: A Margaret Cavendish Reader*, ed. Sylvia Bowerbank and Sara Mendelson (Peterborough, Ontario: Broadview Press, 2000), 143–44.

45. Ibid.

46. *SL*, 25.

47. *SL*, 26.

48. Cavendish, "The Unnatural Tragedy," in *Plays* (London, 1662), 332.

49. Cavendish, "The Second Part of *Bell in Campo*," in *The Convent of Pleasure and Other Plays*, ed. Anne Shaver (Baltimore: Johns Hopkins University Press, 1999), 167.

50. *BW*, 121–23.

51. *PL*, 47.

52. On Cavendish and Hobbes, see Jay Stevenson, "The Mechanist-Vitalist Soul of Margaret Cavendish," *Studies in English Literature* 36 (1996): 527–543; Sara Hutton, "In Dialogue with Thomas Hobbes: Margaret Cavendish's Natural Philosophy," *Women's Writing* 4 (1997): 421–31; and Anna Battigelli, *Margaret Cavendish and the Exiles of the Mind* (Lexington: University of Kentucky Press, 1998), 62–84.

53. Cavendish complains in her *Sociable Letters* that she heard a "Great Scholar, and a Learned Man," probably Hobbes, read her husband's works very poorly while they were at Chatsworth, the Devonshire's country home (185).

54. *PL*, 47.

55. The text of Newcastle's letter is included in *Ideology and Politics on the Eve of Restoration: Newcastle's Advice to Charles II*, transcribed and introduced by Thomas P. Slaughter (Philadelphia: American Philosophical Society, 1984), and James Jacob and Timothy Raylor, "Opera and Obedience: Thomas Hobbes and *A Proposition for Advancement of Moralitie* by Sir William Davenant," *Seventeenth Century* 6 (1991): 215–25, 215–25. Hilda Smith compares Cavendish's and Newcastle's views in " 'A General War amongst the Men.' "

56. Quoted in Steven Shapin and Simon Schaffer, *Leviathan and the Air-Pump: Hobbes, Boyle and the Experimental Life* (Princeton: Princeton University Press, 1985), 284–93.

57. Newcastle, in Slaughter, *Ideology and Politics on the Eve of Restoration*, 224–30.

58. Cavendish, "A Preface to the Reader," in *PL*, n.p.

59. *PL*, 17.

60. Cavendish, "A Preface to the Reader," in *PL*, n.p.

61. Cavendish, "Another Epistle to the Reader," in *PPO 1663*, n.p.

62. Ibid.

63. *PPO 1655*, 29; *PPO 1663*, 73.

64. *PL*, 95.

65. *PL*, 505.

66. *PL*, 96.

67. *PL*, 22.

68. *PL*, 18–19. The passage Cavendish is critiquing is *Leviathan*, 13–14.

69. *PL*, 25. Hobbes's definition of inertia can be found in *Leviathan*, 15.

70. *PL*, 60. Hobbes's doctrine of endeavour or "conatus" is much more complex than Cavendish thought. For Hobbes, endeavour is "motion made through the length of a point, and in an instant or point of time." It is both a dynamic and psychological principle and appears in different forms in many of Hobbes's works (the classic discussion of Hobbes's theory is Frithiof Brandt, *Thomas Hobbes' Mechanical Conception of Nature* [Copenhagen: Levin & Munksgaard, 1928], 293–315). The passage Cavendish is critiquing here comes from *De Corpore*, pt. 4, chap. 25, art. 2 (Hobbes, *The English Works of Thomas Hobbes of Malmesbury*, ed. Sir William Molesworth [London: John Bohn, 1840; repr., Germany: Scientia Verlag Aalen, 1966], 1:391).

71. *PL*, 61.

72. *PL*, 47–48.

73. Ibid.

74. *PL*, 95–96.

75. *PL*, 30. By "imagination," Hobbes means any mental perception. In the full passage in *Leviathan*, chap. 3, Hobbes concluded, "All Fancies are Motions within us, reliques of those made in the Sense" (20). If Hobbes's psychology is correct, Cavendish's whole corpus of work is philosophically problematic.

76. *PL*, 79.

77. Cavendish, preface to *PL*, n.p.

78. *P&F*, 4.

79. *PL*, 38; Hobbes, *Leviathan*, 30. For Hobbes, speech is composed of names, which are signs to designate objects. Human knowledge and communication consist of the comparison and juxtaposition of names. On Hobbes's theory of language, see Douglas Jesseph, "Hobbes and the Method of Natural Science," in *The Cambridge Companion to Hobbes* (Cambridge: Cambridge University Press, 1996), 96–97.

80. *PL*, 40; Hobbes, *Leviathan*, 34.

81. *PL*, 40–41.

82. Keith Thomas, *Man and the Natural World: A History of the Modern Sensibility* (New York: Pantheon Books, 1983), 40.

83. Hobbes, *Leviathan*, 24.

84. Ibid., 89.

85. Ibid., 21.

86. Ibid.

87. *PL*, 43–44.

88. *PL*, 7.

89. Hobbes, "Philosophical Rudiments concerning Government," in *English Works*, 2:8.

90. *PL*, 193. Cavendish's rejection of causal determinism is therefore rooted in her belief in all things being able to know, at least in their own ways.

91. *PL*, 41.

92. *PL*, 41–43.

93. *PL*, 236.

94. Cavendish, "The First Part of *Bell in Campo*," in Shaver, *Convent of Pleasure and Other Plays*, 119.

95. Cavendish, *Orations*, in Bowerbank and Mendelson, *Paper Bodies*, 143.

96. *SL*, 26.

97. "To His Excellency The Lord Marquis of Newcastle," in *SL*, 4–5.

98. *SL*, 37.

99. *SL*, 34.

### 6: THE CHALLENGE OF IMMATERIAL MATTER

1. Cavendish, "A Preface to the Reader," in *PL*, n.p.

2. *SL*, 155.

3. Cavendish, "A Preface to the Reader," in *PL*, n.p.

4. Ibid. Unfortunately, this work has not survived.

5. Ibid.

6. Cavendish, "To the Reader," in *PPO* 1663, n.p.

7. *PL*, 2.

8. *PL*, 12–13.

9. Ibid.

10. Erica Harth, *Cartesian Women: Versions and Subversions of Rational Discourse in the Old Regime* (Ithaca, NY: Cornell University Press, 1992), 3. Harth's discussion is much more nuanced than this quotation would suggest. See also Genevieve Lloyd, *The Man of Reason: "Male" and "Female" in Western Philosophy* (Minneapolis: University of Minnesota Press, 1984).

11. *PL*, 111.

12. *PL*, 98–99.

13. In this chapter, whenever the word *nature* is capitalized, it refers to the personified figure of Nature.

14. Joanna Hodge, "Subject Body and the Exclusion of Women from Philosophy," in *Feminist Perspectives in Philosophy*, ed. M. Griffiths and M. Whitford (Basingstoke, England: Macmillan, 1988), 153

15. Karl Stern, "Descartes," in *Feminist Interpretations of Descartes*," ed. Susan Bordo (University Park: Pennsylvania State University Press, 1999), 46. The different essays in this collection show the degree of scholarly debate about gender and Descartes. The epistemological critique is described in Susan Bordo's own essay, "Selections from the Flight of Objectivity," 64–65.

16. Harth, *Cartesian Women*, 3, 9. Harth adds that the universalizing discourse of Cartesian philosophy was never used by women without some degree of tension: "The writings of the cartésiennes suggest that at the inception of modern rational discourse there were women who offered tentative revisions of the Cartesian model. . . . With hindsight, we begin to understand that the late twentieth-century feminist critique of Cartesian rationalism has a history" (65).

17. On the response of Conway and Elisabeth to Descartes, see Sarah Hutton, "Women Philosophers and the Early Reception of Descartes: Anne Conway and

Princess Elisabeth," in *Receptions of Descartes: Cartesianism and Anti-Cartesianism in Early Modern Europe*, ed. Tad M. Schmalz (London: Routledge, 2005), 3–23, and Sarah Hutton, *Anne Conway: A Woman Philosopher* (Cambridge: Cambridge University Press, 2004), 38, 43–52.

18. Adrien Baillet, *The Life of Monsieur Des-Cartes*, trans. S. R. (London: R. Simpson, 1993), 146.

19. Sir Charles Cavendish to John Pell, March 21/31, 1646, in Noel Malcolm and Jacqueline Stedall, eds., *John Pell (1611–1685) and His Correspondence with Sir Charles Cavendish: The Mental World of an Early Modern Mathematician* (Oxford: Oxford University Press, 2005), 473. Sir Charles's reading of Descartes can be ascertained from other letters in this correspondence (369, 377, 546).

20. *P & F*, 187.

21. We can piece together the etymology of this word through a close reading of its meanings in the seventeenth century. According to the *Oxford English Dictionary*, one meaning of kernel is "a gland or glandular body; a tonsil; a lymphatic gland or ganglion." It is used in this way by Robert Boyle: "That little Kernel in the Brain, call'd by many writers the Conarion" (*The excellency of theology compar'd with natural philosophy (as both are objects of men's study) / discours'd of in a letter to a friend by T.H.R.B.E. . . .; to which are annex'd some occasional thouhts about the excellency and grounds of the mechanical hypothesis* [London, 1674], 140. A conarion at this period was defined as "the pineal gland of the brain (held by Descartes to be the seat of the soul). According to Henry More's usage in *An antidote against atheisme, or, An appeal to the natural faculties of the minde of man, whether there be not a God* (London, 1653), 43, "As for that little sprunt piece of the Braine which they call the *Conarian*, that this should be the very substance whose naturall faculty it is to move it self." Gary Hatfield describes Descartes's use of the pineal gland to explain the physiological relationship of mind and body: "He designated the pineal gland the seat of mind-body interaction, citing a variety of reasons, including the fact that the gland is unitary [as is consciousness], is centrally located, and can be easily moved by the animal spirits" (Gary Hatfield, "Descartes' Physiology and Psychology," in *The Cambridge Companion to Descartes*, ed. John Cottingham [Cambridge: Cambridge University Press, 1992], 349–50).

22. *P & F*, 176.

23. *PL*, 111.

24. René Descartes, *Descartes: Philosophical Letters*, trans. and ed. Anthony Kenny (Minneapolis: University of Minnesota Press, 1970), 182.

25. *PL*, 111–12.

26. *PL*, 107.

27. *PL*, 97.

28. On Descartes's theory of animal automatons, see Leonora Cohen Rosenfield, *From Beast Machine to Man-Machine: Animal Soul in French Letters from Descartes to La Metrie* (New York: Octagon Books, 1968), 197–214, and Stephen Gaukroker, *Descartes' System of Natural Philosophy* (Cambridge: Cambridge University Press, 2002).

29. Descartes to the Marquess of Newcastle, November 23, 1646, in Descartes, *Descartes*, ed. Kenny, 207.

30. *PL*, 114.

31. Descartes to the Marquess of Newcastle, November 23, 1646, in Descartes, *Descartes*, ed. Kenny, 208.

32. *PL*, 111.

33. *PL*, 133.

34. *PL*, 115.

35. *PL*, 101–6.

36. *PL*, 97–98.

37. *PL*, 107. Descartes believed that God put matter in motion and conserved its motions. This motion is a closed, rotating spiral, in which the contacts between bits of matter eventually wear down matter into spherical shapes. Owing to centrifugal force, these spheres form vortices, which are composed of three different sizes of matter: the larger become planets, the smallest and most fluid become suns, and the intervening spaces are filled with medium particles whose interstices are themselves filled with the finest matter. The classic historian of the mechanical philosophy E. J. Dijksterhuis comments, after describing this system, that it, and Descartes's terrestrial physics, are a "marvellously imaginative treatment" (E. J. Dijksterhuis, *The Mechanization of the World Picture* [London: Oxford University Press, 1961], 413). On Descartes's vortex theory, see also Gaukroger, *Descartes' System of Natural Philosophy*, 15–16.

38. *PL*, 135.

39. Samuel I. Mintz, *The Hunting of Leviathan: Seventeenth-Century Reactions to the Materialism and Moral Philosophy of Thomas Hobbes* (Cambridge: Cambridge University Press, 1962), 84–95.

40. *A Collection of Letters and Poems, Written by Several Persons of Honour and Learning, Upon Divers Important Subjects, to the Late Duke and Dutchess of Newcastle* (London, 1678), 90–91; Rupert A. Hall, *Henry More and the Scientific Revolution* (Cambridge: Cambridge University Press, 1997), 146–67.

41. Hall remarks that Henry More was a "strange, idiosyncratic, and very human author" (*Henry More and the Scientific Revolution*, 127).

42. Quoted in E. A. Burtt, *The Metaphysical Foundations of Modern Science* (1932; reissued, New York: Doubleday Anchor Books, 1952), 141. On the Spirit of Nature, see also R. A. Greene, "Henry More and Robert Boyle on the Spirit of Nature," *Journal of the History of Ideas* 23 (1963): 451–74, and Douglas Jesseph, "Mechanism, Skepticism, and Witchcraft: More and Glanvill on the Failures of the Cartesian Philosophy," in Schmalz, *Receptions of Descartes*, 199–217.

43. *PL*, 149–50.

44. *SL*, 25.

45. Merry E. Wiesner, *Women and Gender in Early Modern Europe*, 2nd ed. (Cambridge: Cambridge University Press, 2000), 37. There has been considerable historical debate about the actual legal position of women in English law. There were certainly

some women who gained a degree of independence both legally and fiscally. Nevertheless, the legal position of married women under the common law, according to the doctrine of coverture, viewed women as minors, if not worse, "As the husband was legally responsible for his wife, coverture placed her in the same category as children, wards, lunatics, idiots, and outlaws" (Sara Mendelson and Patricia Crawford, *Women in Early Modern England* [Oxford: Oxford University Press, 2000], 37). See also Wiesner, *Women and Gender*, 35–41, and Cissie Fairchilds, *Women in Early Modern Europe, 1500–1700* (Harlow, England: Pearson Longman, 2007), 280–84.

46. This description is from the Code of Justinian and quoted in Wiesner, *Women and Gender*, 38.

47. Mendelson and Crawford, *Women in Early Modern England*, 58, 52.

48. *TR*, 52.

49. On Cavendish's relationship with Henrietta Maria, see Anna Battigelli, *Margaret Cavendish and the Exiles of the Mind* (Lexington: University of Kentucky Press, 1998), 13–19.

50. Katie Whitaker, *Mad Madge: The Extraordinary Life of Margaret Cavendish, Duchess of Newcastle, the First Woman to Live by Her Pen* (New York: Basic Books, 2002), 12–13.

51. *TR*, 48–49.

52. *PL*, 162–63.

53. *PL*, 154–55.

54. *PL*, 152.

55. *PL*, 164.

56. *PL*, 145.

57. Carolyn Merchant, *The Death of Nature: Women, Ecology and the Scientific Revolution* (San Francisco: Harper & Row, 1980; repr., Harper Collins, 1990).

58. Quoted in ibid., 168. Merchant, however, credits Henry More with a more benign view of nature, arguing that he understood it to possess a vegetative constitution rather than being simply inert (241–42). I don't think that Cavendish understood him in this sense.

59. *PL*, 194–95.

60. *PL*, 145.

61. *PL*, 13–17.

62. *PL*, 221.

63. *PL*, 164.

64. *PL*, 218–19.

65. *PL*, 196.

66. *PL*, 227.

67. *PL*, 219.

68. Ibid.

69. *PL*, 247.

70. *PL*, 242.

71. A. G. Debus, *The English Paracelsians* (London: Oldbourne, 1965).

72. Walter Pagel, *Joan Baptista Van Helmont: Reformer of Science and Medicine* (Cambridge: Cambridge University Press, 1982), 36.

73. *PL*, 238.

74. *PL*, 239, 275.

75. *PL*, 237.

76. On the different meanings of *occult* in the seventeenth century, see Keith Hutchinson, "What Happened to Occult Qualities in the Scientific Revolution?" *Isis* 73 (1982): 233–53.

77. Pagel, *Joan Baptista Van Helmont*, 23.

78. *PL*, 340.

79. *PL*, 300.

80. *PL*, 301.

81. *PL*, 391.

82. *PL*, 277, 335.

83. *PL*, 243.

84. *PL*, 302.

85. Glanvill to Cavendish, in *Letters and Poems*, 104. The date of this letter is not given.

86. *PL*, 298.

87. *PL*, 302.

88. *PL*, 284.

89. *PL*, 278–79.

90. *PL*, 312–13.

91. *PL*, 312.

92. *PL*, 294.

93. Mendelson and Crawford, *Women in Early Modern England*, 47–48.

94. *PL*, 313–14.

95. *PL*, 314.

96. *PL*, 319.

97. *PL*, 280–81.

98. *PL*, 283. Alchemy was very much part of the activity of the members of the Royal Society, including Robert Boyle and Isaac Newton. On Van Helmont's influence, see William R. Newman and Lawrence M. Principe, *Alchemy Tried in the Fire: Starkey, Boyle, and the Fate of Helmontian Chemistry* (Chicago: University of Chicago Press, 2002), and Principe, *The Aspiring Adept: Robert Boyle and His Alchemical Quest, Including Boyle's "Lost" Dialogue on the Transmutation of Metals* (Princeton: Princeton University Press, 1998).

99. *PL*, 490–91.

7: CAVENDISH AGAINST THE EXPERIMENTERS

1. On Cavendish's relationship with Huygens and experimentalism, see Nadine Akkerman and Marguérite Corporaal, "Mad Science beyond Flattery: The

Correspondence of Margaret Cavendish and Constantijn Huygens," *Early Modern Literary Studies* 14 (2004): 2.1–21, and Hilda L. Smith, "Margaret Cavendish and the Microscope as Play," in *Men, Women, and the Birthing of Modern Science*, ed. Judith P. Zinsser (DeKalb: Northern Illinois Press, 2005), 34–47.

2. These recipes appear in British Library, Harley MS 6491. On Newcastle's interest in experimental philosophy, see Lisa T. Sarasohn, "Thomas Hobbes and the Duke of Newcastle: A Study in the Mutuality of Patronage before the Establishment of the Royal Society," *Isis* 90 (1999): 722–24.

3. *SL*, 117. What Cavendish meant by a "glass-ring" is inconclusive. "Chymists," including Boyle, were trying to make malleable glass, so it may be a reference to this activity. I thank Lawrence Principe for this information.

4. Ibid.

5. Cavendish, "To the Reader," in *OEP*, n.p.

6. Cavendish, "The Preface to the Ensuing Treatise," in *OEP*, n.p.

7. Ibid.

8. Frank Whigham, *Ambition and Privilege* (Berkeley: University of California Press, 1984), 130–31; Curtis Brown Watson, *Shakespeare and the Concept of Renaissance Honor* (Princeton: Princeton University Press, 1960), 63.

9. *TR*, 46.

10. "Letter of Mary Evelyn to Ralph Bohun, 1667," in Bowerbank and Mendelson, *Paper Bodies*, 92.

11. *PL*, 4.

12. These dedications are found in *PPO* 1655, *PL*, *GNP*, and *OEP*.

13. Cavendish, "To the Most Famous University of Cambridge," in *OEP*, n.p.

14. *Letters and Poems*, 28.

15. Steven Shapin, "A Scholar and a Gentleman: The Problematic Identity of the Scientific Practitioner in Early Modern England," *History of Science* 29 (1991): 290.

16. Lisa T. Sarasohn, "Who Was Then the Gentleman? Samuel Sorbière, Thomas Hobbes, and the Royal Society," *History of Science* 42 (2004): 1–22.

17. *OEP*, 59–73. The topics she discusses can be found in Robert Boyle, *Experiments and Considerations Touching Colours* (facsimile, 1665 ed., New York: Johnson Reprint Corp., 1964), 6–8, 42, 68–69, 93–132, 495–96.

18. *PL*, 495–96.

19. Robert Boyle, *The Sceptical Chymist* (London: J. M. Dent & Sons; New York: Dutton, 1911), 7. It is possible that Cavendish may have gotten the idea of making "chymists" into ape-men from Boyle, who in this work writes that Paracelsians are "are like apes, if they have some appearance of being rational, are blemished with some absurdity or other, that when they are attentively considered, make them appear ridiculous" (227).

20. Cavendish, "Observations upon the Opinions of some Ancient Philosophers," in *OEP*, 3.

21. Mario Biagioli, "Scientific Revolution, Social Bricolage, and Etiquette," in *The Scientific Revolution in National Context*, ed. Roy Porter and Mikulas Teich (Cambridge: Cambridge University Press), 32–35. Steven Shapin and Simon Shaffer

argue that in order to avoid dissension and "enthusiasm" the Royal Society privileged the "matter-of-fact" as an authoritative alternative to the various dogmatic claims of natural philosophers in the pre-Restoration period (Shapin and Shaffer, *Leviathan and the Air-Pump: Hobbes, Boyle and the Experimental Life* [Princeton: Princeton University Press, 1985], 72–80).

22. Cavendish, "Further Observations, in *OEP*, 4.

23. Cavendish, "The Preface to the Ensuing Treatise," in *OEP*, n.p.

24. Cavendish, "To the Most Famous University of Cambridge," in *PL*, n.p.

25. Michael Hunter, *Establishing the New Science: The Experience of the Early Royal Society* (Woodbridge, Suffolk, England: Boydell Press, 1989), 167, 171.

26. Cavendish, "To the Reader," in *OEP*, n.p.

27. Ibid.

28. Ibid.

29. The *Oxford English Dictionary* defines a mountebank as someone who "uses entertainments to attract a crowd of potential customers" and who "falsely claims knowledge of some skill in some matter."

30. *OEP*, 11.

31. *OEP*, 10–11.

32. *OEP*, 101.

33. Ibid.

34. *OEP*, 101–2. Sylvia Bowerbank discusses Nature's role as a "Trickster," who makes the efforts of man to understand her "a laughable project. Her [Cavendish's] style of natural philosophy is to ridicule such efforts and to imitate the wily ways of nature, in her shifts, her doublings, her tricks, and contradictions" (Bowerbank, *Speaking for Nature: Women and Ecologies of Early Modern England* [Baltimore: Johns Hopkins University Press, 2004]. 73).

35. *OEP*, 102.

36. Robert Boyle, *Some Considerations Touching the Usefulnesse of Experimental Naturall Philosophy* (London, 1663), 19.

37. *OEP*, 6.

38. *WO*, 13.

39. Cavendish, "Further Observations," in *OEP*, 24.

40. Ibid., 78–79.

41. Ibid., 6–7.

42. *OEP*, 32.

43. *OEP*, 8–9.

44. Cavendish, "Further Observations," in *OEP*, 12–13.

45. *OEP*, 86.

46. *OEP*, 162.

47. Cavendish, "Further Observations," in *OEP*, 54.

48. *OEP*, 11.

49. Cavendish, "Further Observations," in *OEP*, 12.

50. *OEP*, 101–3.

51. Cavendish, "Further Observations," in *OEP*, 14–15.

52. *OEP*, 141–42.

53. *OEP*, 66.

54. Cavendish, "Further Observations," in *OEP*, 15–16.

55. Cavendish, "Some Observations of the Opinions of some Ancient Philosophers," in *OEP*, 2.

56. Rebecca Merrens, "A Nature of 'Infinite Sense and Reason': Margaret Cavendish's Natural Philosophy and the 'Noise' of a Feminized Nature," *Women's Studies* (1996): 2, 430.

57. *OEP*, 24.

58. Cavendish, "To the Reader," in *BW*, n.p.

59. *BW*, 13.

60. *BW*, 1–5.

61. *BW*, 15.

62. *BW*, 16.

63. Cavendish, "The Second Part," in *BW*, 32.

64. *BW*, 22, 26.

65. *BW*, 27.

66. *BW*, 28.

67. Ibid.

68. Ibid.

69. *BW*, 29–30.

70. *BW*, 31–32.

71. *BW*, 56.

72. *OEP*, 23.

73. *PL*, 283.

74. Boyle, *Sceptical Chymist*, 227.

75. On the many meanings of apes in the seventeenth and eighteenth century, see Susan Wiseman, "Monstrous Perfectibility: Ape-Human Transformations in Hobbes, Bulwar, Tyson," in *At the Borders of the Human: Beasts, Bodies and Natural Philosophy in the Early Modern Period*, ed. Erica Fudge, Ruth Gilbert, and Susan Wiseman (Hampshire, England: Palgrave, 2002), 215–38. On the meaning of animals in the new science, see Erica Fudge, "Calling Creatures by Their Real Names: Bacon, the New Science and the Beast in Man," in Fudge, Gilbert, and Wiseman, *At the Borders of the Human*, 91–109.

76. *PL*, 292.

77. *BW*, 46–47.

78. *BW*, 48.

79. *BW*, 45.

80. *NP*, 309–10.

81. Cavendish, "To the Universities," in *PPO* 1655, n.p.

82. *BW*, 41–42.

83. *OEP*, 38.

84. Keith Thomas, *Man and the Natural World: A History of the Modern Sensibility* (New York: Pantheon Books, 1983), 88.

85. *PL*, 198, 133.

86. *BW*, 43.

87. Robert Hooke, *Micrographia: or, Some Physiological Descriptions of Minute Bodies made by Magnifying Glasses. With Observations and Inquiries thereupon* (London, 1665; repr., Brussels: Culture et Civilisation, 1966), 190.

88. "Further Observations," in *OEP*, 41–42.

89. *BW*, 121–22.

90. The control of knowledge production was something advocated by the Duke of Newcastle and his circle, including Thomas Hobbes and the poet William Davenant. See James Jacob and Timothy Raylor, "Opera and Obedience: Thomas Hobbes and *A Proposition for Advancement of Moralitie* by Sir William Davenant," *Seventeenth Century* 6 (1991): 215–25.

91. *BW*, bk. 2, 12–14.

92. Ibid., 14–23.

93. *BW*, 94.

94. *BW*, 97.

95. *BW*, 100.

96. *BW*, 101–2.

97. *BW*, 122.


## 8: MATERIAL REGENERATIONS

1. Cavendish, "Convent of Pleasure," in *The Convent of Pleasures and Other Plays*, ed. Anne Shaver (Baltimore: Johns Hopkins University Press, 1999), 226.

2. *GNP*, 270, 279.

3. The plays are "The Sociable Companions, or the Female Wits," "The Presence," "The Bridals," The Convent of Pleasure," and "A Piece of a Play," all in *Plays Never Before Printed* (London, 1668).

4. By 1668, Newcastle was collaborating with John Dryden on plays. Samuel Pepys thought Newcastle's play "The Humourous Lovers," which was produced in London during the same trip when Cavendish visited the Royal Society, was written by the duchess. He was not impressed; he described it as "the most silly thing that ever came upon a stage." Nevertheless, he went to see it again and reiterated his view that the play was "the most ridiculous thing that ever was wrote, but yet she and her lord mightily pleased with it" (quoted in Geoffrey Trease, *Portrait of a Cavalier, William Cavendish, First Duke of Newcastle* [New York: Taplinger, 1979], 197).

5. Margaret Cavendish, Duchess of Newcastle, *Observations upon experimental philosophy to which is added The description of a new blazing world / written by the thrice noble, illustrious, and excellent princesse, the Duchess of Newcastle* (London, 1668); *Orations of divers sorts, accommodated to divers places. VVritten by the thrice noble, illustrious, and excellent princess, the Dutchess of Newcastle* (London, 1668); *Poems, or Several Fancies In Verse: with the Animal Parliament, In Prose* (London, 1668).

6. Cavendish, "To all Noble and Worthy Ladies," in *BW* 1668, n.p.

7. Quoted in Tease, *Portrait of a Cavalier*, 196.

8. See the Essay on Sources for a discussion of this question.

9. Alexandra G. Bennett, "Fantastic Realism: Margaret Cavendish and the Possibilities of Drama," in *Authorial Conquests: Essays on Genre in the Writings of Margaret Cavendish*, ed. Line Cottegnies and Nancy Weitz (London: Associated Presses, 2003), 179–94; Susan Wiseman, "Gender and Status in Dramatic Discourse: Margaret Cavendish, Duchess of Newcastle," in *Women, Writing, History, 1640–1740*, ed. Isobbel Grundy and Susan Wiseman (Athens: University of Georgia Press, 1992), 159–77; Judith Peacock, "Writing for the Brain and Writing for the Boards: The Producibility of Cavendish's Dramatic Texts," and Julie Sanders, "'The Closet Opened': A Reconstruction of 'Private' Space in the Writings of Margaret Cavendish," both in *A Princely Brave Woman: Essays on Margaret Cavendish, Duchess of Newcastle*, ed. Stephen Clucas (Aldershot, England: Ashgate, 2003), 87–108, 127–42.

10. Cavendish, "A Piece of a Play," in *Plays Never Before Printed*, 5–7.

11. Ibid., 1.

12. Ibid., 9.

13. Cavendish, "The Presence," in ibid., 24.

14. Ibid.

15. *Oxford English Dictionary.*

16. Ibid.

17. John Dryden, *The Works of John Dryden*, vol. 7: *Poems, 1697–1700* (Berkeley: University of California Press, 2002), 502.

18. Cavendish, "Convent of Pleasure," in *Paper Bodies: A Margaret Cavendish Reader*, ed. Sylvia Bowerbank and Sara Mendelson (Peterborough, Ontario: Broadview Press, 2000), 99.

19. Cavendish, *Orations of Divers Sorts, Accommodated to Divers Places* (London, 1662), 187.

20. Cavendish, "Convent of Pleasure," in Bowerbank and Mendelson, *Paper Bodies*, 101–2.

21. Cavendish, "*Orations*," 188–89.

22. Cavendish, "Convent of Pleasure," in Bowerbank and Mendelson, *Paper Bodies*, 99.

23. Ibid., 100.

24. Cavendish, "To the Reader," in *OEP*, n.p.

25. *GNP*, 10–11.

26. *P & F*, 23.

27. Cavendish, "To the Universities in Europe," in *GNP*, n.p.

28. Ibid.

29. *GNP*, 3.

30. *GNP*, 2.

31. *GNP*, 100–101.

32. *GNP*, 177.

33. *GNP*, 24.

34. *P & F*, 45.

35. *PF*, 62.

36. *PPO* 1655, 97–98.

37. *GNP*, 236.

38. *GNP*, 247–48.

39. *GNP*, 258.

40. *GNP*, 261.

41. According to Catholic theology, the immaterial soul immediately after death receives a "Particular Judgment," where it is sent immediately to Heaven, Hell, or (until recently) Purgatory. At the General Resurrection, the soul is united with the body and is forever damned or saved. The difference about the corporeal nature of the soul, of course, separated Cavendish from the orthodox version of this doctrine (*The Catholic Encyclopedia* [New York: Universal Knowledge Foundation, 1909], 6:620).

42. *GNP*, 254.

43. *GNP*, 245.

44. *GNP*, 254.

45. *GNP*, 263. These ideas were first mentioned in Cavendish's *Orations*, in *Margaret Cavendish: Political Writings*, ed. Susan James (Cambridge: Cambridge University Press, 2003), 227–28.

46. *GNP*, 265–66.

47. Whitaker argues that Cavendish made this disclaimer because she was "concerned to avoid any appearance of heresy or irreligion" (Katie Whitaker, *Mad Madge: The Extraordinary Life of Margaret Cavendish, Duchess of Newcastle, the First Woman to Live by Her Pen* [New York: Basic Books, 2002], 320).

48. *GNP*, 277.

49. *GNP*, 279.

50. *GNP*, 10–11.

51. *GNP*, 279.

52. *GNP*, 278–79.

53. *GNP*, 279.

54. *GNP*, 282.

55. *GNP*, 288, 283.

56. *GNP*, 284.

57. *GNP*, 283.

58. *GNP*, 270.

59. *GNP*, 291–92.

60. *GNP*, 101.

61. Whether Cavendish is a mortalist, like Hobbes and Milton, who believed that the body died until resurrected with the soul at the Second Coming, is unclear here. On mortalism, see Christopher Hill, "Irreligion in the 'Puritan' Revolution," in *Radical Religion in the English Revolution*, ed. J. F. McGregor and B. Reay (Oxford: Oxford University Press, 1984), 199–204.

62. *GNP*, 293.

63. *GNP*, 296.

64. *GNP*, 296. Cavendish discusses the impossibility of chemical transformation in *OEP*, 72.

65. *GNP*, 308.
66. *GNP*, 298.
67. *GNP*, 294.
68. *GNP*, 296.
69. *GNP*, 304.
70. *GNP*, 310–11.

### CONCLUSION: DOES CAVENDISH MATTER?

1. Robert Darnton, *The Great Cat Massacre and Other Episodes in French Cultural History* (New York: Basic Books, 1984), 78.

2. Cavendish, "The Epilogue to the Reader," in *OEP*, incorrectly paginated as 121–22.

3. Abraham Cowley, "To the Royal Society," in *The History of the Royal-Society of London, For the Improving of Natural Knowledge* (London, 1667), n.p.

4. Quoted in Dorothy Stimson, "The Critical Years of the Royal Society, 1672–1703," *Journal of the History of Medicine* (Summer 1947): 295. Wotton was referring specifically to *The Virtuoso* (1676) by Thomas Shadwell, which was dedicated to William Cavendish.

5. Ursula Le Guin is one of the foremost modern writers of science fiction. Her novel *The Left Hand of Darkness* (New York: Ace Books, 1976) depicts a world whose inhabitants are hermaphrodites, alternatively becoming male and female. On Mary Shelley's *Frankenstein* and its use of the monstrous, see Stephen Bann, ed., *Frankenstein, Creation and Monstrosity* (London: Reaktion Books, 1994).

6. *Letters and Poems in Honour of the Incomparable Princess, Margaret, Dutchess of Newcastle*, 166.

# Essay on Sources

Margaret Cavendish's increasing prominence in studies of seventeenth-century literature and science has resulted in an explosion of works about her and her ideas, mostly written by students of English literature. This study aims to interpret Cavendish in the context of the scientific revolution (an increasingly problematic category), the history of gender, and the history of political thought.

## PRIMARY SOURCES
### Works by Margaret Cavendish Cited in This Book

*Philosophicall Fancies*. London, 1653.
*Poems, and Fancies*. London, 1653.
*Philosophical and Physical Opinions*. London, 1655; 2nd ed., 1663.
*The Worlds Olio*. London, 1655.
*Natures Pictures*. London, 1656; 2d ed., 1671.
*A True Relation of my Birth, Breeding, and Life*. London, 1656.
*Plays*. London, 1662.
*Philosophical Letters*. London, 1664.
*Sociable Letters*. London, 1664.
*Description of a New World, Called the Blazing World*. London, 1666.
*Observations upon Experimental Philosophy*. London, 1666.
*Life of William Cavendishe*. London, 1667.
*Grounds of Natural Philosophy*. London, 1668.
*Plays Never Before Printed*. London, 1668.

### Other Seventeenth-Century Works Cited in This Book

Boyle, Robert. *The excellency of theology compar'd with natural philosophy (as both are objects of men's study) / discours'd of in a letter to a friend by T.H.R.B.E. . . . ; to which are annex'd some occasional thoughts about the excellency and grounds of the mechanical hypothesis*. London, 1674.
Charleton, Walter. *The Darkness of Atheism dispelled by the Light of Nature, A Physico-theologicall Treatise*. London, 1652.

———. *Physiologia Epicuro-Gassendi-Charletoniana, or a Fabrick of Science Natural Upon the Hypothesis of Atoms, Founded by Epicurus, Repaired by Petrus Gassendus. Augmeted by Walter Charleton.* London, 1654. Reprint, New York: Johnson Reprint, 1966.

———. *A ternary of Paradoxes the Magnetick Cure of Wounds, Nativity of Tartar in Wine, Image of God in Man / written originally by Joh. Bapt. Van Helmont and translated, illustrated and amplified by Walter Charleton.* London, 1650.

*A Collection of Letters and Poems, Written by Several Persons of Honour and Learning, Upon Divers Important Subjects, to the Late Duke and Dutchess of Newcastle.* London, 1678.

*Letters and Poems in Honour of the Incomparable Princess, Margaret, Dutchess of Newcastle.* London, 1676.

More, Henry. *An Antidote against Atheisme, or, An appeal to the natural faculties of the minde of man, whether there be not a God.* London, 1653.

Newcastle, Willam Cavendish, Duke of. *La Méthode de invention nouvelle de dresser les chevaux.* Anvers, 1657–58.

———. *A New Method and Extraordinary Invention to Dress Horses, and Work Them According to Nature; as Also to Perfect Nature by the Subtlety of Art; Which Was Never Found Out But by the Thrice Noble, High and Puissant Prince, William Cavendish, Duke of Newcastle.* London, 1667.

Sprat, Thomas. *The History of the Royal-Society of London, For the Improving of Natural Knowledge.* London, 1667.

Winstanley, Gerard. *The True Leveller Standard Advanced.* London, 1649.

## Modern Editions and Archival Collections

In 2001, Cambridge University Press published a new edition of *Observations upon Experimental Philosophy* (1668). This volume is the only one of Cavendish's expressly scientific works to be republished since the seventeenth century, except for a small facsimile reproduction of *Grounds of Natural Philosophy* (West Cornwall, CT: Locust Hill Press, 1996). Unfortunately, the Cambridge edition does not include *Blazing World*, an absence that obscures the continuity of Cavendish's critique of experimentalism. Modern editions of *Blazing World* are available in *The Blazing World and Other Writings* (London: Pickering, 1992; rpt., London: Penguin, 1994), with an excellent introduction by Kate Lilley, the volume's editor, and in *Paper Bodies: A Margaret Cavendish Reader*, edited by Sylvia Bowerbank and Sara Mendelson (Peterborough, Ontario: Broadview Press, 2000). This well-chosen collection of Cavendish's works devotes a section to her natural philosophy and also gives comparable texts by Francis Bacon and Aphra Behn; it includes Cavendish's autobiography, *A True Relation of My Birth, Breeding, and Life. Blazing World* and *Orations of Divers Sorts* are reprinted in *Margaret Cavendish: Political Writings*, edited by Susan James (Cambridge: Cambridge University Press, 2003).

Additional context for Cavendish's life and work can be found in James Fitzmaurice's edition of *Sociable Letters* (New York: Garland, 1997; 2nd ed., Ontario: Broadview, 2004), including a valuable introduction and short biography of the duchess. Cavendish's *Life* of William Cavendish, and a copy of her autobiography, also appeared in an early twentieth-century edition, entitled *The Life of the (1st) Duke of Newcastle and Other Writings by Margaret Duchess* (London: J. M. Dent & Sons, 1916). Several of Cavendish's plays are included in *The Convent of Pleasure and Other Plays* (Baltimore: Johns Hopkins University Press, 1999), with an informative introduction by the volume's editor, Anne Shaver.

Noel Malcolm's edited volumes *The Correspondence of Thomas Hobbes* (Oxford: Clarendon Press, 1994) and, with Jacqueline Stedall, *John Pell (1611–1685) and His Correspondence with Sir Charles Cavendish: The Mental World of an Early Modern Mathematician* (Oxford: Oxford University Press, 2005) are invaluable in any study of mid-seventeenth-century English science and make available archival material that previously had to be read on site. Volume 8 of Robert Latham and William Matthew's edition of *The Diary of Samuel Pepys* (Berkeley: University of California Press, 1974) includes the description of Cavendish's visit to the Royal Society and other details of her London tour in 1667 (242–43).

The major archival collections relating to the Cavendishes are located at the British Library (Additional MSS 4278, 4280; Harlean MS 6988) and the University of Nottingham (Portland MSS Pw1, Pw5).

## SELECTED SECONDARY SOURCES

The analysis of Cavendish's role in early modern thought can be divided into three areas: science, gender, and politics. The discussion of Cavendish's scientific ideas began with Virginia Woolf's well-known comment in *The Common Reader*: "Under the pressure of such vast structures, her natural gift, the fresh and delicate fancy which had led her in her first volume to write charmingly of Queen Mab and fairyland, was crushed out of existence." Another evaluation is less familiar, but more telling: "There is something noble and Quixotic and high-spirited, as well as crackbrained and bird-witted about her. . . . She has the freakishness of an elf, the irresponsibility of some non-human creature, its heartlessness, and its charm" (*The Common Reader* [New York: Houghton Mifflin/Harcourt, 2002], 79, 77).

The idea that Cavendish was eccentric and strange continues to influence some interpretations of 'her role in science, particularly the accounts of Cavendish's visit to the Royal Society in 1667. Samuel I. Mintz, in the first scholarly account of the visit, "The Duchess of Newcastle's Visit to the Royal Society," *Journal of English and Germanic Philology* 51 (1952): 168–76, pronounced Cavendish's reaction to the experiments she saw as "childlike" (176). A recent popular biography of Cavendish by Katie Whitaker, *Mad Madge: The Extraordinary Life of Margaret Cavendish, Duchess of Newcastle, the First Woman to Live by Her Pen* (New York: Basic Books, 2002), suggests that the duchess was "overcome" during her visit (298–300), and Frances Harris, in "Living in the Neighbourhood of Science: Mary Evelyn, Margaret Cavendish and the Greshamites," in *Women, Science and Medicine, 1500–1700: Mothers and Sisters of the Royal*

*Society,* edited by Judith P. Zinsser (Phoenix Hills, England: Sutton, 1997), argues that Cavendish "muted her criticism of the new science" after her visit (210).

Other scholars interpret the visit differently. Anna Battigelli, in her intellectual biography *Margaret Cavendish and the Exiles of the Mind* (Lexington: University of Kentucky Press, 1998), recognizes that the "admiration" Cavendish professed while viewing the society's entertainment was feigned (112). Emma Rees, in her excellent account of Cavendish's use of different genres, *Margaret Cavendish: Gender, Genre, Exile* (Manchester: Manchester University Press, 2003), credits her with agency, arguing that she used eccentricity as a cover in order to attack other natural philosophers. Sarah Hutton, in "Anne Conway, Margaret Cavendish and Seventeenth-Century Scientific Thought," in Zinsser, *Women, Science and Medicine, 1500–1700*, notes Cavendish's willingness to accept ridicule to bolster her claim to originality, and in her study *Anne Conway: A Woman Philosopher* (Cambridge: Cambridge University Press, 2004), Hutton contrasts Cavendish's public style with Conway's more private intellectual endeavors. Sylvia Bowerbank, in *Speaking for Nature: Women and Ecologies of Early Modern England* (Baltimore: Johns Hopkins University Press, 2004), analyzes the visit within the context of Cavendish's natural philosophy and sees her professed admiration as a "fitting response to the agency of Nature herself " (64).

Other scholars also treat Cavendish as a consequential figure who contributed to the scientific revolution as both a critic and a theorist. My 1985 article "A Science Turned Upside Down: Feminism and the Natural Philosophy of Margaret Cavendish," *Huntingdon Library Quarterly* 47 (1984): 289–307, associated Cavendish with the atomistic tradition of Pierre Gassendi and his circle. Londa Schiebinger discusses Cavendish's theories at length in *The Mind Has No Sex?* (Cambridge, MA: Harvard University Press, 1989) and argues that her scientific achievements should be taken seriously. Susan James's "The Philosophical Innovations of Margaret Cavendish," in the *British Journal for the History of Philosophy* 7 (1992): 219–44, focuses on Cavendish's natural philosophy, its originality, and its ties to other intellectual traditions and concludes that Cavendish, to the extent she was influenced by anyone, is closest to Hobbes (244). Sarah Hutton, in a very important discussion of Cavendish's natural philosophy, "In Dialogue with Thomas Hobbes: Margaret Cavendish's Natural Philosophy," *Women's Writing* 4 (1997): 421–32, argues that Cavendish is best understood as a Hobbist and that her inclusion in accounts of the scientific revolution allows it to be understood in its own terms, rather than through the prism of later definitions of science.

Other commentators link Cavendish with diverse philosophic traditions. Sujata Iyengar, in "Royalist, Romancist, Racialist: Rank, Gender, and Race in the Science and Fiction of Margaret Cavendish," in *ELH* 69 (2002): 649–72, argues that Cavendish was indebted to Cartesianism (653). Jacqueline Broad, in *Women Philosophers of the Seventeenth Century* (Cambridge: Cambridge University Press, 2007), also associates Cavendish with "popular Cartesianism" (40) but recognizes that she rejected Cartesian dualism in part because of her sympathy with animals.

The strongest case for Cavendish's debt to another philosophy is Eileen O'Neill's argument that Cavendish was influenced by the Stoic doctrine of *pneuma*, or divine spirit, in her natural philosophy, which O'Neill develops in her introduction to her edition of *Observations*. Cavendish's concept of innate matter is indeed very close to the Stoic concept of *pneuma*, or soul, but Cavendish does not devote much time to Stoicism in her text, and so it is difficult to make an argument based on the connection between this ancient philosophy and her philosophy. Her ethical thought is clearly closer to Epicurean hedonism, and her materialism reflects the hylozoistic tendencies of Continental and English atomism. Stephen Clucas discusses Cavendish's vitalistic materialism within the context of seventeenth-century atomism in a very important article, "The Atomism of the Cavendish Circle: A Reappraisal," *Seventeenth Century* 9 (1994): 247–73.

Clucas has also edited an excellent collection of essays, *A Princely Brave Woman: Essays on Margaret Cavendish, Duchess of Newcastle* (Aldershot, England: Ashgate, 2003), which includes a section on Cavendish's natural philosophy. *Authorial Conquests: Essays on Genre in the Writings of Margaret Cavendish*, edited by Line Cottegnies and Nancy Weitz (London: Associated University Presses, 2003), has a section devoted to Cavendish's scientific ideas. A special edition of the journal *In-between: Essays and Studies in Literary Criticism* 9 (2000) contains several interesting essays devoted to Cavendish's natural philosophy in the context of gender and science.

Other works on Cavendish and science include Nadine Akkerman and Marguérite Corporaal, "Mad Science beyond Flattery: The Correspondence of Margaret Cavendish and Constantijn Huygens," *Early Modern Literary Studies* 14 (2004): 1–21, and Hilda Smith, "Margaret Cavendish and the Microscope as Play," in *Men, Women, and the Birthing of Modern Science*, edited by Judith P. Zinsser (DeKalb: Northern Illinois Press, 2005). Akkerman, Corporaal, and Smith all trace Cavendish's early participation in experimental culture. In a more linguistic analysis, Elizabeth Spiller, "Reading through Galileo's Telescope: Margaret Cavendish and the Experience of Reading," *Renaissance Quarterly* 53 (2000): 192–221, argues that Cavendish reimagined perception through the mediated experience of "seeing" through a telescope.

Some recent interpreters of Cavendish's scientific ideas see her natural philosophy as intrinsically feminist or protofeminist, in both its ontology and epistemology. Eve Keller, in "Producing Petty Gods: Margaret Cavendish's Critique of Experimental Science," *ELH* 64 (1997): 447–71, claims "a rather startling similarity between Cavendish's position and a post-Kuhnian and even a proto-feminist critique of the rational basis of mechanical science" as part of her argument for the importance of gender in examining early modern science and its critics (251–52). Carolyn Merchant, in her controversial *The Death of Nature: Women, Ecology, and the Scientific Revolution* (San Francisco: Harper & Row, 1980), concludes her brief summary of Cavendish's natural philosophy by asserting that Cavendish "presented one of the earliest explicitly feminist perspectives on science" (272). Deborah Boyle, in "Margaret Cavendish's Nonfeminist Natural Philosophy," *Configurations* 12 (2004): 195–227, challenges the feminist interpretation, arguing that Cavendish's skeptical epistemology was not a philosophy of knowledge particularly associated with women, nor was Cavendish's organic natural philosophy peculiar to her gender. Although not directly discussing Cavendish, the debate about using gender as a category of analysis in the history of science continues in the "Focus" section of *Isis* 97 (2006): 487–495, in which Katharine Park defends Merchant and argues that considering gender allows for a new synthesis more inclusive of the diverse currents present in the early modern investigation of nature.

Recent commentators also try to link Cavendish's vitalistic materialism to various political traditions. John Rogers, in *The Matter of Revolution: Science, Poetry, and Politics in the Age of Milton* (Ithaca, NY: Cornell University Press, 1996), argues that Cavendish was a protoliberal who with some discomfort "structured as an unabashedly liberal system, a system of disseminated sovereignty devised quite specifically to counter the authoritarian organization of the leading theories of her day" (197). Rogers's argument is powerful but rests on the contention that vitalism was necessarily "liberal" because the particles of matter are endowed with agency and autonomy. Cavendish would have been astonished to discover that her idea comes closest to the Puritan ideal of rule by "the Godly few"(199).

Most modern analyses of Cavendish's political thought differ on whether she favored a monarchical system of government, in either its royalist or Hobbesian versions. Hilda Smith, "'A General War amongst the Men . . . But None amongst the Women': Political Differences between Margaret and William Newcastle," in *Politics and the Political Imagination in Later Stuart Britain* (Rochester, NY: Rochester University Press, 1997), argues for the difficulty of assessing Cavendish's political leanings but concludes that "Cavendish speaks outside the range of acceptable royalist and Anglican viewpoints" (153). Iyengar, in "Royalist, Romancist, Racialist," classifies Cavendish and her husband as "ardent Royalists" but argues that fiction allowed her to imagine a world where she could question gender and racial hierarchies without compromising her own belief and reliance on class in the real world (650–51). Deborah Boyle, in "Fame, Virtue, and Government: Margaret Cavendish on Ethics and Politics," *Journal of the History of Ideas* 67 (2006): 251–89, argues forcefully, and I believe correctly, that Cavendish's first political and moral imperative was the construction of a government that would secure peace. While she enumerates various possible views of government, ultimately Cavendish endorsed absolutism as the most effective means of maintaining peace (282).

I am indebted to broader studies of intellectual and scientific developments in early modern Europe. Cavendish was part of the skeptical movement first described by Richard Popkin in *The History of Scepticism from Erasmus to Descartes* (New York: Harper & Row, 1964) and by Barbara Shapiro in *Probability and Certainty in Seventeenth-Century England: A Study of the Relationships between Natural Science, Religion, History, Law, and Literature* (Princeton: Princeton University Press, 1983). Both of these studies seek to understand the social context of the development of ideas, as does the deservedly classic *Leviathan and the Air-Pump: Hobbes, Boyle and the Experimental Life* (Princeton: Princeton University Press, 1985) by Steven Shapin and Simon Schafer. Mario Biagioli's *Galileo: Courtier* (Chicago: University of Chicago Press, 1993) inspires my work on patronage. Michael Hunter's work on the Royal Society, particularly *Establishing the New Science: The Experience of the Early Royal Society* (Woodbridge, Suffolk, England: Boydell Press, 1989) and *Science and*

*Society in Restoration England* (Cambridge: Cambridge University Press, 1981), is essential for understanding the early years of this institution.

Cavendish's adversarial stance toward the Royal Society reflected social attitudes that valued wit, dialogue, and humor in the investigation of nature and welcomed imagination as not antithetical to reason but as a legitimate aspect of human cognition. J. B. Shank's essay on Bernard de Fontenelle, "Neither Natural Philosophy, nor Science, nor Literature," in Zinsser, *Men, Women, and the Birthing of Modern Science,* shows how imagination and fancy slowly became associated with a popular and female audience and separated from the activities pursued by the Académie Royale des Sciences. My discussion of wit, women, and wonder is also informed by Lorraine Daston and Katie Park's *Wonders and the Order of Nature, 1150–1750* (New York: Zone Books, 1998) and Mary Baine Campbell's *Wonder and Science: Imagining Other Worlds in Early Modern Europe* (Ithaca, NY: Cornell University Press, 1999).

The idea that nature could be a source of amusement as well as edification in early modern Europe owes much to Paula Findlen's *Possessing Nature* (Berkeley: University of California Press, 1994) and the many articles she has written about the practice of natural philosophy in Italy. The broadening of natural philosophy to include natural history reflects Keith Thomas's magisterial study of the role of animals in early modern Europe, *Man and the Natural World: A History of the Modern Sensibility* (New York: Pantheon Books, 1983). I profited greatly from Erica Fudge's sophisticated analysis of the role of animals in English culture and ideas, *Perceiving Animals: Humans and Beasts in Early Modern English Culture* (Hampshire, England: Palgrave, 1999) and *Brutal Reasoning: Animals, Rationality and Humanity in Early Modern England* (Ithaca, NY: Cornell University Press, 2006). The role of popular culture during this period is best analyzed by Barry Reay in his *Popular Cultures in England, 1550–1750* (London: Longman, 1998) and in the articles in Tim Harris, ed., *Popular Culture in England, c. 1500–1850* (New York: St. Martin's Press, 1995).

# Index

absolutism, 101–2, 105–10, 115, 119, 165, 170

actions, 90, 92, 101, 217n55; brain, with senses, 39, 41, 44, 48, 50, 133–34; in nature, 136, 139

admiration, 31–32

agency, 62, 122, 217n55, 217n59

air, 43, 95, 167, 181

alchemy, 1–2, 6, 144, 147, 149, 199n3; matter, 55–57, 60–61, 94, 142; particulate, 51, 212n26

angels/Angel, 23, 98, 117, 141, 143, 170

animals: automatonism, 133–34; drama imagery, 13, 122–23, 176–79; experimentation, 23, 31, 164–65; matter, 43, 53, 68, 181, 187–88; rationality, 11–12, 72–73, 85, 121–22, 130, 170, 194; soul, 53, 60–61, 97–98

animate beings/matter, 18, 47, 49, 59, 104, 106, 132

antipathy, 56, 58, 63, 65, 84, 211n7

ape-men, 1, 152, 159, 165, 167, 228n19

Aristotle, 2, 8–9, 45, 56, 72, 120, 156, 211n7

art/Art, 10–11, 13, 147, 159–63, 166–67

astronomy, 68, 73, 95–96, 158, 160, 165–66, 181

atheism, 36, 47, 53, 62, 67, 85–86, 136, 140–41, 184

atomism, 12, 192, 212n26; autonomy, 43–45, 57, 64–65, 81, 104; Cavendish's, 34, 38–49, 55, 68–69, 97, 162; Epicurean, 2, 34–37, 41, 46–48, 51–52, 62, 65–68, 85, 103, 179; fairies, 35–37, 39, 48–55, 58–62, 131–32, 169, 181; politics, 101–4, 219n7; religion vs., 35–36, 45–46, 62, 66, 82, 86–87

authority: attack on, 16–17, 22, 32, 82, 155; Cavendish's challenge, 6, 12–13, 36–39, 138–39, 192–93

autonomy: individual, 104, 119, 137–38, 194, 225n45; matter, 43–45, 54–55, 57, 64–65, 80–81, 104

Bacon, Francis, 2, 6, 19, 139, 163

bear-men, 1, 152, 165–67

beast-men, 1, 13, 73, 95, 121, 176–79, 186, 191; Blazing World, 152, 164–72

beings: kinds, 23, 49, 103, 122, 142–43; other world, 95–96

bestiality, 97, 146, 160; female, 7–8, 13, 82, 123–24, 177, 191

*Blazing World*, 1, 3, 10, 191, 195; experimental philosophy, 149, 151–54, 157, 161–72; material regeneration, 174–77, 185–88; natural philosophy, 18, 29, 32, 203n13; politics, 106, 113–14, 124, 165

body: mind/soul connection, 87–88, 91, 97, 104, 129–34, 216n45, 224n21; motion, 59, 70–71, 130–33; regeneration vs. resurrection, 42–45, 95, 173, 179–88, 218n73, 233n41

Boyle, Robert, 2, 10–11, 16, 28–29, 31–32, 154, 158, 162, 167, 228n19

brain: actions, with senses, 39, 41, 44, 48, 50, 133–34; materialism, 63, 131, 183–84, 189; order vs. disorder, 71–72, 110

cause and effect, 89–90, 158–59, 161, 165, 171

Cavendish, Margaret: appeal to universities, 79–80, 152–57, 180, 187; appearance,